CCRP Publication Series

This is a continuation in the series of publications produced by the Center for Advanced Concepts and Technology (ACT), which was created as a “skunk works” with funding provided by the CCRP under the auspices of the Assistant Secretary of Defense (NII). This program has demonstrated the importance of having a research program focused on the national security implications of the Information Age. It develops the theoretical foundations to provide DoD with information superiority and highlights the importance of active outreach and dissemination initiatives designed to acquaint senior military personnel and civilians with these emerging issues. The CCRP Publication Series is a key element of this effort.

Check our Web site for the latest CCRP activities and publications.

www.dodccrp.org

DoD Command and Control Research Program

Assistant Secretary of Defense (NII)
&
Chief Information Officer
Mr. John P. Stenbit

Principal Deputy Assistant Secretary of Defense (NII)
Dr. Linton Wells, II

Special Assistant to the ASD(NII)
&
Director, Research and Strategic Planning
Dr. David S. Alberts

Library of Congress Cataloging-in-Publication Data

Moffat, James, 1948-
Complexity theory and network centric warfare / James Moffat.
p. cm. -- (Information age transformation series)
ISBN 1-893723-11-9 (pbk. : alk. paper)
1. War--Mathematical models. 2. Information warfare. 3. Computational complexity. I. Title. II. Series.
U21.2 M584 2003
355.4'8--dc21

2003000365

September 2003

Information Age Transformation Series

Complexity Theory and Network Centric Warfare

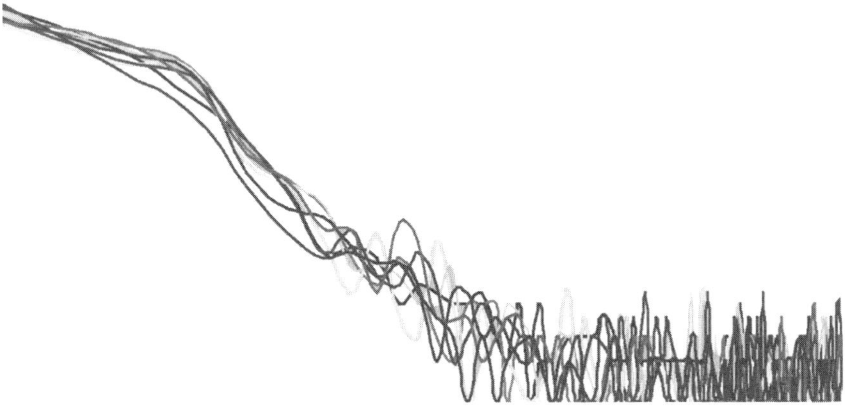

James Moffat

CCRP Publication Series

This book is dedicated to my wife Jacqueline and my children,
Louise and Katherine.

NOTES TO THE READER

Although I use the term *Complexity Theory* as if it was a coherent body of scientific theory, this area of research is in fact still both young and evolving. I use it therefore as a shorthand term to cover a number of areas, each with its own distinct heritage. Broadly, it covers fractal structures, nonlinear dynamical systems, and models of self-organisation and self-organised criticality.

The research on which this book is based could not have been carried out without the help of a number of other people. Their contributions are, I hope, suitably acknowledged in the text. I would like to thank particularly Walter Perry (RAND), Susan Witty (Dstl), David Rowland, and Maurice Passman for their contributions. I am also most grateful to Professor Henrik Jensen for contributing the Foreword.

Table of Contents

LIST OF FIGURES

LIST OF TABLES

FOREWORD

For the last couple of decades, attempts have been made to develop some general understanding, and ultimately a theory, of systems that consist of many interacting components and many hierarchical layers. It is common to call these systems *complex* because it is impossible to reduce the overall behaviour of the system to a set of properties characterising the individual components. Interaction is able to produce properties at the collective level that are simply not present when the components are considered individually. As an example, one may think of mutuality and collaboration in ecology. The function of any ecosystem depends crucially on mutual benefits between the different species present. One example is the relation between legumes, such as peas and beans, and their associated nitrogen-fixing bacteria: the bacteria collects nitrogen for the legume, which in turn produces carbohydrates and other organic material for the bacteria. Clearly this crucial arrangement cannot be studied by focusing on, say, the legume and neglecting the bacteria; the ecological function emerges first when the different components are brought together and interaction is taken into account.

Another important feature of complex systems is their sensitivity to even small perturbations. The same action is found to lead to a very broad range of responses, making it exceedingly difficult to perform prediction or to develop any type of experience of a "typical scenario." This must necessarily lead to great caution: do not expect what worked last time to work this time. The situation is exacerbated since real systems (ecological or social) undergo adaptation. This implies that the response to a given strategy most likely makes the strategy redundant. An example is the effect of using the same type of antibiotic against a given type of bacteria. Evolution soon ensures that the bacteria develop resistance and make the specific type of antibiotic useless. That complex systems adapt and change their properties fundamentally as a result of the intrinsic dynamics of the system is clearly extremely important. Nevertheless, for the sake of simplicity adaptation is often neglected in model studies. Sometimes assuming the existence of a stationary state might be justified (e.g., if one is interested in "toy" models of the flow of granular material under a controlled steady input of grains). But if one is dealing with more complex situations such as in ecology, and even more when considering social and political systems, ignoring adaptation is very likely to lead to erroneous conclusions.

We know from studies of Self-Organised Critical models, which the present book alludes to (for more see P. Bak, *How Nature Works*, Oxford University Press, 1997 and H.J. Jensen, *Self-Organized Criticality*, Cambridge University Press, 1998), that the correlations and general behaviour exhibited by these model systems are entirely determined by the assumed boundary conditions or the applied drive. The lesson to be learned from this is that complex systems cannot be studied independently of their surroundings. Understanding the behaviour of a complex system necessitates a simultaneous understanding of the environment of the system. In model studies, one assumes often that the surroundings can be represented by one or the other type of "noise," but this is just a trick that allows one to proceed with the analysis without understanding the full system under consideration. It is very important to appreciate that the "drive" or the "noise" are equally crucial to the understanding, as is the analysis of the "system" itself. One should bear in mind that the separation into system, drive, noise, surroundings, etc. is rather arbitrary and is far from representing a complete analysis.

From these considerations, we see that it is vitally important to consider warfare as a complex system that is linked and interacts (in a coevolving way) with the surrounding socio-economical and political context. From that perspective, the present book is a "work in progress" and a preliminary first step along the road in helping to analyse and structure these difficult and serious issues. Forgetting that war and warfare are an intimate part of a much larger complex system will lead to incomplete and even dangerously incorrect conclusions. Applying the approach of Complexity Theory to warfare leads one to the self-consistent realisation that warfare will have to be analysed in its larger context. Further work will need to examine how coevolution across the entire network of military, socioeconomical, and political interactions leads firstly to emergent effects at higher levels, and of equal importance how such effects lead to coevolution at the higher level. It will also be important to consider the robustness of such networks, and their vulnerability to damage.

Henrik Jeldtoft Jensen
Professor of Mathematical Physics
Department of Mathematics
Imperial College, London

CHAPTER 1

COMPLEXITY IN NATURAL AND ECONOMIC SYSTEMS[1]

In this chapter we consider some of the key ideas of Complexity Theory as applied to natural systems. Having established these key ideas, in Chapter 2 we begin to see how these ideas map across, in a broad conceptual sense, to the dynamics of conflict in an Information Age environment. In later chapters, we will look in more detail at the evidence and the type of modelling that emerges from this conceptual connection. In an organisational context, we argue that complexity provides an explanatory framework of interrelationships, both metaphorically and analogously, of how individuals and military organisations interact, relate, and evolve

[1]The contribution of Dr Maurice Passman to this chapter is gratefully acknowledged.

within a larger "ecosystem." Complexity explains why interventions may have unanticipated consequences, but also explains how combat effects follow from these consequences. The intricate interrelationships of elements within a complex system give rise to multiple chains of dependencies. Change happens in the context of this intricate intertwining at all scales. We become aware of change only when a different pattern becomes discernible. But before change at a macro level can be seen, it is taking place at many micro levels simultaneously. Hence, microcomponent interaction and change leads to macrosystem evolution.

In a previous book,[2] we considered some of the issues to be addressed at the political/military level as a consequence of such emergent behaviour and the "resultant likelihood of complex and unexpected interactions, arising from previously unexpected sources." The diligent reader is directed to that work for further discussion of these issues. Here, we simply wish to add weight to the points made by Professor Jensen in his Foreword to this present work. In all that follows, the recursion of the process up to this political/military level must be kept in mind, and will be one of the key areas of future research.

Our lead is taken from current military doctrinal thought both in the UK and in the United States (particularly the U.S. Marine Corps [1]). The Chief Analyst of the UK Defence Science and Technology Laboratory (Dstl), Roger Forder, makes the following point in his discussion of the future of defence analysis [2]:

---

[2]Chapter 1 of: Moffat J (2002). *Command and Control in the Information Age.* The Stationery Office. London, UK.

> *One effect of the human element in conflict situations is to bring a degree of complexity into the situation such that the emergent behaviour of the system as a whole is extremely difficult to predict from the characteristics and relationships of the system elements. Detailed simulation, using agent-based approaches, is always possible but the highly situation-specific results that it provides may offer little general understanding for carrying forward into robust conclusions of practical significance. Usable theories of complexity, which would allow understanding of emergent behaviour rather than merely its observation, would therefore have a great deal to offer to some of the central problems facing defence analysis. Indeed they might well be the single most desirable theoretical development that we should seek over the next few years.*

A similar thought was aired at a Royal United Services Institute (RUSI) conference on future Intelligence, Surveillance, Target Acquisition, and Reconnaisance (ISTAR) [3]. Vice Admiral Cebrowski,[3] U.S. Navy, centred his keynote address on Network Centric Warfare (NCW) as the capstone concept for the U.S. Navy after Next. He described it in terms of the achievement of Information Superiority with characteristics of gross asymmetries and a diversity of "players." These ideas are explicitly derived in his description from the new physics of nonlinearity, complexity, and chaos as exemplified by the Santa Fe Institute corpus of ideas. The relationship between complexity and "information-based" warfare is (as he described it) less deterministic and more emergent; less focused on the physical, and more behavioural; less focused on things, and more on relationships. Command and Control (C2)

[3]Now head of the Office of Force Transformation, The Pentagon, U.S. DoD.

emphasises speed, sharing, and decentralisation. In summary, ADM Cebrowski defined NCW as the robust networking of well-informed, geographically dispersed forces.

Where do these ideas come from?

In looking at where these key ideas of complexity come from, let us make a start by considering systems and their behaviour in the natural world. In the classical view, such physical or biological processes are reducible to a few fundamental interactions. This leads to the idea that under well-defined conditions, a system governed by a given set of laws will follow a unique course (like the planets of the solar system). Moreover, a slight change in the cause will likewise produce a slight change in the effects (i.e., the system is *linear* in nature). Recently, an increasing amount of experimental data challenging this idea has become available and this imposes a new attitude concerning the description of nature. In natural systems, under appropriate conditions, a multitude of self-organisation phenomena on a macroscopic scale (a scale of an order of magnitude larger than the range of fundamental interactions) in the form of spatial or temporal patterns may be generated.

A Simple Example

To illustrate this, let us consider a simple thermodynamic thought experiment. Imagine a layer of fluid limited by two horizontal parallel plates whose lateral dimensions are much longer than the width of the layer, as shown in Figure 1.1.

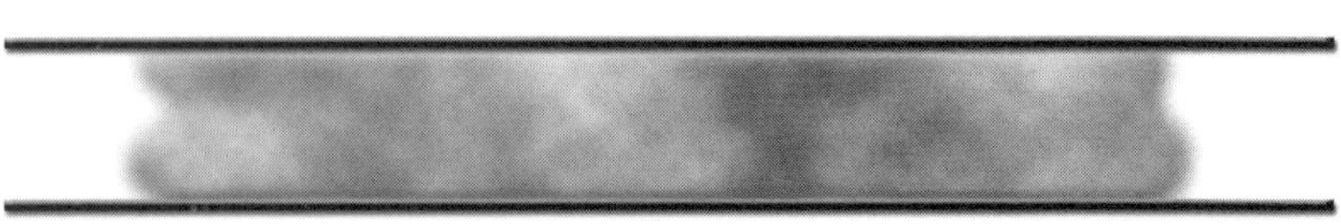

Figure 1.1: Two Horizontal Plates Containing a Layer of Fluid

Left to reach equilibrium, the fluid will rapidly tend to a homogeneous state that is statistically identical. The homogeneity of this system extends to all of its properties, particularly to its temperature, which will be the same at all parts of the fluid and equal to the temperature of the limiting plates or, alternatively, to the temperature of the "external" world. All of these properties are characteristic of a system in a particular state, *the state of equilibrium*, for which there is neither bulk motion nor temperature difference with the outside world. What occurs if, for example, the temperature on a small section on one of the plates is temporarily perturbed? At equilibrium, this temperature perturbation has no influence, since the temperature rapidly becomes uniform again and equal to its initial value. In other words, the perturbation dies out; the system keeps no track of it. Such a state is said to be *asymptotically stable.*

From the standpoint of a very small observer inside the system, not only does the homogeneity of the fluid make it impossible for the development of an intrinsic concept of space, but also the stability of the state of equilibrium eventually makes all time instances identical. It is therefore impossible for this observer to develop an intrinsic conception of correlation or coincidence (e.g., things happening at the same time, or in the same place). We can increase the complexity of the system by, for instance, heating the fluid layer from below. In doing this, we communicate energy to the system in the form of heat. Moreover, as the temperature of the lower plate is now higher than the upper, the equilibrium condition is violated. In other words, by applying *external constraints* to the system, we do not permit the system to reach equilibrium. The presence of an external constraint therefore implies energy *flux* and vice-versa.

Now suppose at first that the constraint is weak, i.e. the change in temperature, ΔT, is small. The system will again adopt a simple and unique state in which the only active process is a transfer of heat from the lower to the upper plate, from which heat is lost to the external world. The only difference from the state of equilibrium is that temperature, and consequently density and pressure, are no longer uniform. They vary from warm regions to cold regions in an approximately linear fashion. This phenomenon is known as *thermal conduction.* In this new state that the system has reached in response to a constraint, stability will prevail again and the behaviour will eventually be as simple as at equilibrium. However, by removing the system from equilibrium further and further, through an increase in ΔT, we observe that suddenly, at a value of ΔT that we will call *critical,* matter begins to perform a bulk movement. Moreover, this movement is far from random; the fluid is structured in a series of small structures known as Benard cells.

Owing to thermal expansion, the fluid closer to the lower plate is characterised by a lower density than that nearer the upper plate. This gives rise to a gradient of density that opposes the force of gravity. This configuration is thus potentially unstable. Consider a small volume of the fluid near the lower plate. Imagine that this volume is displaced upward by a perturbation. This volume, now in a colder and hence denser region, will experience an upward Archimedes force, amplifying the ascending movement further. If, on the other hand, a small droplet initially close to the upper plate is displaced downward, it will penetrate an environment of low density and the Archimedes force will tend to amplify the initial descent. The fluid thus generates the observed currents. The stabilising effect of viscosity, which generates an internal friction opposing movement, counteracts the destabilising effects. This, and thermal conduction, which tends to average any temperature difference between the displaced droplet and its environment, explains why currents do not appear as soon as ΔT is not strictly zero.

Benard cells also show the complexity of movement. The cells unfold along the horizontal axis, adopting successively right-handed or left-handed rotation. Our very small observer can now locate his position in space by considering the rotation of the cell he occupies and by counting the number of cells he passes through. The emergence of this notion of space is known as *symmetry breaking*. When ΔT is below the critical value, the homogeneity of the fluid in the horizontal direction renders its different parts independent of each other. In contrast, beyond the threshold, it is as if each volume element is watching the behaviour of its neighbours and is taking this into account in order to play its role adequately and to participate in the overall pattern. This suggests the existence of *correlations* of statistically reproducible rate relations between distant parts of the system. The characteristic space dimension of a Benard cell is in the millimetre range, whereas the characteristic space scale of the intermolecular forces is in the Angstrom range. That large numbers of particles can behave in a coherent fashion at this long range, despite random thermal motion, is one of the principal properties characteristic of such *self-organisation* and emergent complex behaviour.

This experiment is reproducible; the same convection patterns will appear at the same threshold value and the process is subject to a strict determinism. However, the *direction* of the rotation of the cells is unpredictable. The form of the particular perturbation that prevails at the moment of the experiment will decide whether a given cell is right- or left-handed. When the constraint is sufficiently strong, several solutions are possible for the same parameter values and chance alone will decide which of these solutions is realised. In this way, the system has been perturbed from a state of equilibrium or near-equilibrium to a state of *self-organisation*, with a number of possible modes of behaviour.

What happens to the Benard cell system when the thermal constraint is increased beyond this first threshold? For some range of values the Benard cells will be maintained globally, but some of their specific characteristics will be modified. Further constraint induces the system to move beyond another critical point and turbulence is witnessed. Note that all of these critical behaviours are different from the phase changes we normally associate with closed thermodynamic systems. The reason behind this is that a *nonequilibrium constraint* is being applied. For example, the dendritic structure associated with snowflakes has nothing to do with the structure of the underlying ice-crystal lattice. The scale, size, and spacing of the emergent structure is of an order of magnitude larger.

To summarise, nonequilibrium has enabled the system to transform part of the energy communicated from the environment into an ordered behaviour of a new type: the *dissipative structure*. This regime is characterised by *symmetry breaking*, *multiple modes of behaviour*, and *correlation*. Such a system is called "open" since it is open to the effect of energy or information flowing into and out of the system. It is also called "dissipative" because of such energy flows, and the resultant dissipation of energy.

Open, Dissipative Structures

Consider then a system embedded in an environment with which it communicates through the exchange of certain properties that we shall call *fluxes*. We now know to call this an open system (in contrast to a closed or isolated system). Figure 1.2 is a schematic representation of such an open system, communicating with the environment through the exchange of such properties as mass, energy, or information. The rate of amount transported per unit surface is the flux of the corresponding property across the system. In our simple example above, the amount of heating is the flux of energy into the system.

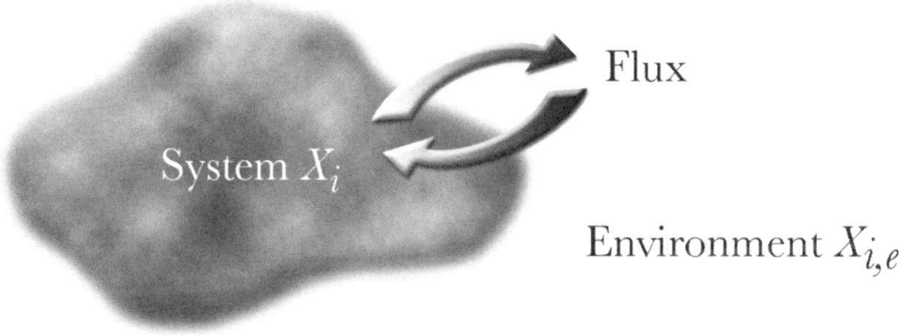

Figure 1.2: Schematic of an Open System

As a result of these exchanges, the variables describing the instantaneous state of the system, $\{X_i\}$, vary in time and attain values typically different from those characterising the state of the environment $\{X_{i,e}\}$. Whatever the detailed interpretation of X_i might be, the evolution of the system under consideration may be described in the following general form:

Rate of change of the system state = function of (system state and control variables).

Thus we have:

$$\frac{dX_i}{dt} = F_i(X_1,...X_n;\lambda_1,...\lambda_m), \quad (i = 1,....n)$$

where F_i denote the laws concerning the rate of change of the system, and $\lambda_1,...\lambda_m$ are a set of parameters present in the problem, which can be modified by the external world. These quantities are known as *control parameters*. Under certain conditions, this relation will have a single solution that minimises some measure of negative utility (which we call a *loss function*). This solution is then a unique *optimal control for the system*. For each time t, it defines an optimal value for the settings of the control parameters $\lambda_1,...\lambda_m$.

An Example

If we think of a guided missile attempting to manoeuvre towards a target, the measure of *loss* is the miss distance relative to the aim point. The control parameters are the settings for the missile fins at a given time t. For simple forms of *linear* guidance (e.g., early forms of laser-guided bombs), this leads to what is called *bang-bang* control, where the missile fins "bang" from one extreme setting to another in order to keep the missile on course. The Appendix goes into this in more detail and shows that such solutions correspond to maximising or minimising a Hamiltonian function. This is due to Pontryagin's maximum principle. Applied to a *linear* control system, this maximum principle leads to the solution of bang-bang control.

A characteristic feature of many of the systems encountered in nature, however, is that the F's are complicated *nonlinear* functions of the X's. The equations of evolution of this type of system should then admit, under certain conditions, several solutions (rather than just the one optimal solution) since a multiplicity of solutions is the most typical feature of a nonlinear equation. Our assumption will be that these solutions represent the various *modes of behaviour* of the underlying system.

Modes of Behaviour in Nonlinear Systems

How do these different modes of behaviour arise in such a nonlinear system? Thermodynamic equilibrium is characterised by *detailed balance*, i.e.

Probability of a "direct" process = Probability of a "reverse" process

We can understand that in such a state, any attempt at diversification and self-organisation will be smeared out immediately: equilibrium is a state of full homeostasis, characterised both by uniqueness and robust stability properties. Our aim is to extend our ideas of equilibrium to the nonequilibrium dynamics of an open, dissipative system.

The most useful view of equilibrium is as follows. We represent the evolution of the system in a space spanned by the state variables (*phase space*). An instantaneous state of the system is thus represented in phase space by a point. As the system evolves over time, a succession of such states is produced, giving rise to a curve in phase space, which is called a *phase space trajectory*. In a dissipative dynamical system, as time progresses, the phase space trajectory will tend to a limit representative of the regime reached by the system when all transients die out. We call this regime the *attractor*. The attractor representing an equilibrium position is unique and describes a time-independent situation. This gives a phase space point towards which all possible histories converge monotonically. The state of equilibrium is therefore a *universal point attractor*. The goal of *self-organisation* is thus the search for new attractors that arise when a system is driven away from its state of equilibrium. For example, consider the type of coupled chemical system studied by Ilya Prigogine [4] (for which he received the Nobel Prize in Chemistry):

$$A + B \underset{k_2}{\overset{k_1}{\rightleftarrows}} 3X$$

$$X \underset{k_4}{\overset{k_3}{\rightleftarrows}} B$$

Here the concentration $[x]$ of product X is taken to be the only state variable, being understood that A and B are continuously supplied from or removed to the outside to maintain fixed concentrations. At equilibrium, detailed balance implies that the rate equation is:

$$k_1[a][x]^2 = k_2[x]^3$$

$$k_3[x] = k_4[b]$$

These relations fix the equilibrium value $[x_e]$ of x uniquely and impose a condition on the concentrations of constituents A and B:

$$[x_e] = \frac{k_4[b_e]}{k_3} = \frac{k_1[a_e]}{k_2}$$

$$\frac{[b_e]}{[a_e]} = \frac{k_1 k_3}{k_2 k_4}$$

In a dissipative stationary state far from equilibrium, it is not necessary for each individual reaction to balance in both directions. Cancelling the overall effect of the two forward reactions by that of the backward reactions is sufficient, and this yields a cubic equation:

$$-k_2[x_s]^3 + k_1[a][x_s]^2 + k_3[x_s] - k_4[b] = 0\,.$$

This equation may have up to three solutions (hence three possible modes of behaviour) for certain values of [*a*] and [*b*]. It may therefore be said that nonequilibrium reveals the potentialities hidden in the nonlinearities, which remain "dormant" at or near equilibrium.

The monotonic character of the approach to the state of equilibrium implies that the evolution laws of the system should obey, *in the neighbourhood of equilibrium*, very particular conditions. Introducing the deviations of X_i from the equilibrium values, $X_{i,e}$:

$$x_i = X_i - X_{i,e}$$

the evolution of $\{x_i\}$ near equilibrium can then be written:

$$\frac{dx_i}{dt} = -\sum_j \Gamma_{i,j} \left(\frac{\partial \Phi}{\partial x_j} \right).$$

Φ is a thermodynamic potential taking its minimum at equilibrium and $\{\Gamma_{i,j}\}$ is a symmetric matrix. This symmetry can be traced back to the property of detailed balance or the property of the invariance of the equilibrium state to time reversal. We can see from this that the system response is essentially linear near to equilibrium (i.e., small changes lead to small effects).

The Far-from-Equilibrium State

The search for a generalised thermodynamic potential Φ in the nonlinear range well away from the equilibrium state has attracted a great deal of attention, but these efforts have, so far, not made much progress. Typically, therefore, beyond the linear domain for such irreversible processes, the above control equations are expected to break down. A first consequence is that the steady state point *attractor* of the system (i.e., the point to which the system state moves as the system evolves over time), extrapolating the state of equilibrium as the distance from equilibrium is increased, can now be approached through damped oscillations. This behaviour heralds a still more interesting possibility in which the oscillation eventually becomes sustained. Topologically, this implies the emergence of a new one-dimensional attractor in phase space. The point attractor for the state of the system is essentially replaced by a circle. In the limit, the system moves endlessly around this circle, which is thus known as a *limit cycle.*

By allowing the intrinsic nonlinearity to be manifested in the regime of detailed balance, nonequilibrium can also lead to the coexistence of multiple attractors in state space. The state space can then be carved up into a set of basins. Each of these corresponds to the set of states that, if the system were to start from there, would evolve to a particular attractor. These are known as the *basins of attraction*. The ridges separating these basins of attraction are called *separatrices*. The coexistence of multiple attractors constitutes the natural mode of systems capable of showing adapted behaviour and of performing regulatory tasks. We would thus expect to see the system staying within one basin of attraction (corresponding to resistance to change) and then at some point switching between different attractors (corresponding to a change in the long-term mode of behaviour) as we further vary the initial state of the system. The existence of one-dimensional attractors (points and circles) suggests the possibility of higher dimensional attracting objects in phase space. These model multiperiodic and chaotic behaviour, which is observed under appropriate experimental conditions.

Nonequilibrium phenomena show a variety of behaviours and therefore correspond to the movements of the system towards different attractors. The simplest mechanism to depict this is the bifurcation diagram, where a single control parameter (the thermal gradient in the case of Benard cells) affects the dynamics of the system.

Bifurcations and Universality Limits

Consider a system described by a set of evolution laws of the form of the equation:

$$\frac{dX_i}{dt} = F_i\left(X_1,...X_n;\lambda_1,...\lambda_m\right), \quad (i=1,....n)$$

where, as before, F_i are the rate laws, and λ_i are the control parameters. In a typical natural phenomenon, the number of variables n is expected to be very high. This will considerably complicate the search for all possible solutions. However, suppose that by experiment we know one of the solutions. By a standard method, known as *Linear Stability Analysis*, we can then determine the parameter values for which this solution regarded as the reference state switches from asymptotic stability to instability.

Robustness and Linear Stability Analysis

Stability or "robustness to change" is essentially determined by the response of the system to perturbations. It is therefore natural to transform the dynamical laws into a form in which the perturbations appear explicitly:

$$X_i(t) = X_{i,s} + x_i(t)$$

and

$$\frac{dx_i}{dt} = F_i(\{X_{i,s} + x_i\},\lambda) - F_i(\{X_{i,s}\},\lambda)\,.$$

These equations are homogeneous in the sense that the right hand side vanishes if all $x_i = 0$. Expanding, we may write:

$$\frac{dx_i}{dt} = \sum_j L_{ij}(\lambda)x_j + h_i(\{x_j\},\lambda) \quad i = 1,2,\ldots..n$$

where L_{ij} are the coefficients of the linear part and h_i are the nonlinear part. It is assumed that the asymptotic stability of the reference state (i.e., $X = X_s$ or $x = 0$) of the system is identical to that of the linearised part:

$$\frac{dx_i}{dt} = \sum_j L_{ij}(\lambda)x_j \quad i = 1,2,\ldots..n .$$

This is reasonable provided that the perturbation is not too large and the system is "well behaved."

In general, a multivariate system gives rise to a wide spectrum of values for the rate of change of perturbations. For a unique control variable $\lambda = \lambda_c$, two cases can be distinguished. First, the perturbations are nonoscillatory and the bifurcations (i.e., alternative modes of behaviour) will correspond to steady-state point attractors, or secondly, the perturbations are oscillatory and the bifurcations will correspond to time-periodic solutions in the form of limit cycles. More intricate solutions can also be envisaged leading to secondary, tertiary, or higher order interactions; however, the complete stable unfolding of the problem remains an open question. It is left to other methods, such as explicit simulation using cellular automata (i.e., agent-based simulation) to attempt to solve the problem of describing the dynamic behaviour of nonequilibrium systems. The use of such a simulation approach can be described as "experimental mathematics."

Self-Organisation in Nature–An Example

Consider an ecosystem consisting of a large number of interacting species, each evolving in response to the environment created by the rest of the ecosystem (i.e., each species is *coevolving*). Such a system consists of many components that interact through some kind of exchange of forces or information. In addition to the internal interactions, the system may be driven by some external force–natural selection in this case. The system will now evolve over time under the influence of the external driving forces and the internal interactions. What happens when we observe such a system? Is there some simplifying mechanism that produces a typical behaviour shared by large classes of such systems? The mechanism, it turns out, is that of *clustering*.

A simple cellular automaton model of such an ecosystem is given by the Bak-Sneppen evolution model. It is an example of the experimental mathematics we described earlier, as a way of analysing the coevolution of such a complex system over time. A description of this model of coevolution within an ecosystem was given in the last chapter of [5] as part of an introduction to ideas from complexity mathematics and the development of mathematical "metamodels" of future Information Age conflict. For ease of reference, some of that description is repeated here.

The Bak-Sneppen Evolution Model

In this automaton model, we have a *d*-dimensional lattice (Figure 1.3) and random numbers f_i drawn without replacement from the interval $[0,1]$ occupy the lattice sites. At each update step, the extremal site (that is, the one with the smallest value of f_i) is chosen, and then it and its $2d$ immediate neighbours are assigned new random numbers. As a model of evolution, the values f_i correspond to "fitness" values. Changing both the site and neighbouring sites captures the process of local coevolution. It follows from [6] that the set of such active sites is a fractal in space-time (see particularly Figures 1 and 28 of that reference).

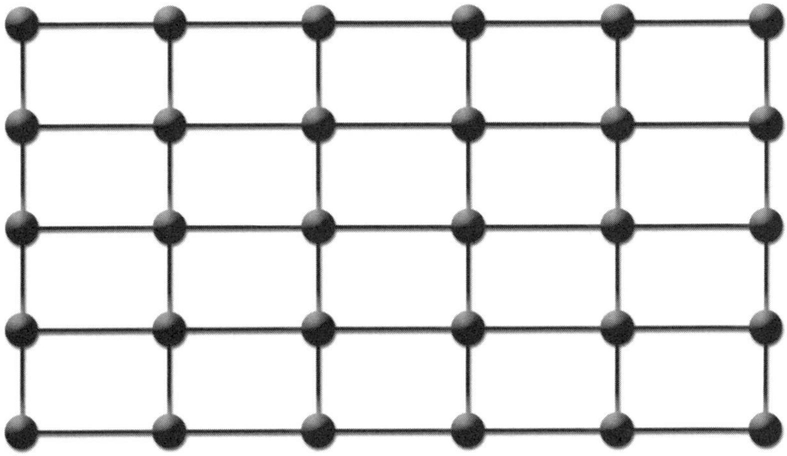

Figure 1.3: The Lattice of Species Interactions in a Model Ecosystem

As described by [5, Chapter 6], the approach to the critical attractor of the process (at which avalanches/clusters of all sizes are possible) is controlled by the "gap equation":

$$\frac{dG(s)}{ds} = \frac{1-G(s)}{L^d \langle S \rangle_{G(s)}}$$

where $G(s)$ is the maximum extremal value $f_i(s)$ at time s, corresponding to the gap opened up between the existing state of the system and the zero or ground state, L is the linear size of the lattice, and $\langle S \rangle_{G(s)}$ is the average avalanche size at timestep s. (An avalanche or cluster consists of a set of extremal values f_i, each of which is a neighbour of the previous extremal value. The size of an avalanche is the number of timesteps for which this process continues.) Schematically, we thus have the following picture of the process (Figure 1.4):

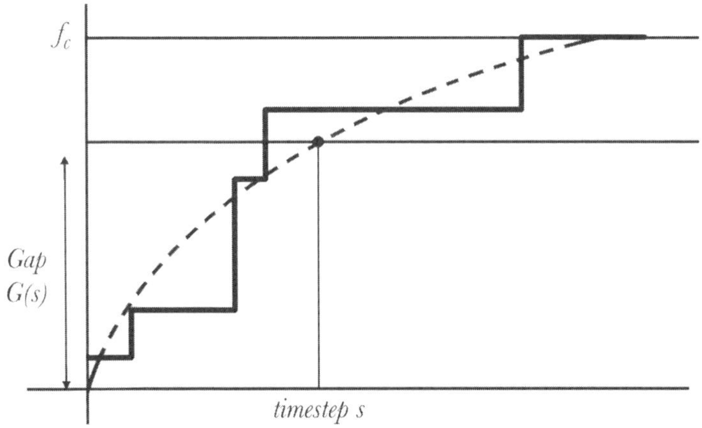

Figure 1.4: Movement of the Ecosystem towards a Self-Organised Critical Point

Figure 1.4 shows how the equation (the hatched line) approximates the self-organised movement of the system, via a series of avalanches/clusters towards the critical attractor of the system at which the system has optimal flexibility (in the sense that clusters of all sizes can be created). This critical point corresponds to a fitness value f_c. At this point, there are no fitness values below this critical value and a flat distribution of fitness values in the range from f_c to 1.0. This is in complete contrast to the behaviour of a closed system such as an ideal gas in an isolated container, where the gas evolves from (for example) being partitioned in part of the container to the equilibrium state where it is spread equally throughout.

Such critical systems are of particular scientific interest. Systems in critical states do not have any characteristic scale and may therefore exhibit the full range of behavioural characteristics within the particular system restraints. This means that systems at the point of criticality are in a position of optimal flexibility in some sense, as we have noted. It could thus be argued [5] that one of the requirements of military command is to so arrange things that the forces collaborate locally and thus self-organise into this optimal state.

From the previous equation, the rate of change of the gap is inversely proportional to the average avalanche size:

$$\dot{G}(s) \propto \frac{1}{\langle S \rangle_{G(s)}} .$$

Thus at the critical value $f_c, \langle S \rangle_{G(s)} \rightarrow \infty$ and near to criticality, the average avalanche size satisfies the scaling law:

$$\langle S \rangle \sqcup (f_c - f_i)^{-\gamma}$$

for some exponent γ (as has been confirmed by experiment).

Before the critical point is reached, at some time t, if f_0 is the smallest random number on the lattice in the evolution model, then random numbers created at the next time-step will only continue the avalanche process if they are smaller than f_0. Thus the value f_0 can be viewed as the branching probability of a random process over time. This will give information on the avalanche size. If f_0 = branching probability, then for larger f_0 we have larger avalanches. We thus assume [6] a scaling relation of the form:

$$P(S, f_0) = S^{-\tau} g(S(f_c - f_0)^{1/\sigma})$$

to describe the probability distribution of avalanches/clusters of size S corresponding to an extremal value f_0. Such a relation has also been confirmed by simulation experiments. In particular, when f_0 equals f_c, the critical value, the probability of a cluster of size S is given by $P(S, f_c) \sqcup S^{-\tau}$. We can think of the function g as a 'cutoff' corresponding to the fact that the avalanches created will have a finite size. As we move to the critical state, this cutoff dies away. We shall call τ the *clustering exponent* for this coevolving ecosystem. In the context of manoeuvre warfare, this describes the statistics of local clustering/collaboration at a transient point f_0 heading towards the critical attractor value f_c. The parameters τ and σ are model dependent and g is our "scaling function." Note again that the average size of the f_0 avalanche diverges as $f_0 \rightarrow f_c$, i.e.

$$\langle S \rangle \sqcup (f_c - f_i)^{-\gamma}.$$

If we mark each of the minimal sites on the lattice as it is identified as an extremal value f_0, then the set of marks generated over time forms a fractal in space-time [6] as we have already noted. Cuts of this fractal in the space direction at a given time identify the site that is "active" (i.e., chosen as the minimal site) at that time. Cuts in the time direction produce a fractal time series. In Chapter 3, we show that the time series of casualties in conflict has the characteristics, in some cases, of such a fractal time series. One interpretation of such effects is that it is the dynamics of local clustering (by one side) that is leading to casualties to the other side. This can also be shown to occur in cellular automata models of conflict such as the ISAAC model produced by the U.S. Marine Corps Combat Development Centre, as discussed in Chapter 6 of reference [5].

Self-Organised Criticality

In a paper published in 1987, Bak, Tang and Weisenfeld [7] first proposed the hypothesis that a system consisting of many interacting constituents may exhibit, in certain cases, a specific general emergent behaviour characteristic of the system. Bak described the behaviour of this type of system by the term *self-organised criticality* (SOC). Self-organisation has for many years been used to describe the ability of certain nonequilibrium systems to develop structure and patterns. The word *criticality* has a very precise meaning in thermodynamics. It is used in connection with phase transitions. At all temperatures other than the transition temperature, perturbations of the system will only locally influence system components. At the critical temperature, the perturbation affects the whole system, even though only the nearest neighbour system components interact directly. The system becomes critical in the sense that all of the members of the entire system influence each other. For the example ecosystem above, the system self-organises itself into the critical state corresponding to this ability of the entire system to be influenced through the propagation of local coevolution influences and the resultant clusters/avalanches of species that coevolution created.

Separation of Internal and External Timescales

Such a self-organised dynamical state requires the separation of the external and internal timescales. For example, the stress in the earth's crust, built up over a period of time, is of a different scale to the subsequent earthquake lasting merely minutes or seconds. The force applied to an individual tectonic plate must overcome a threshold in order to produce an earthquake. This means that the plate may exist in a multitude of intermediate states—*metastable states*—on the way to criticality. Among all of the metastable states, some are of particular importance. These states are *marginally* stable; a slight disturbance may lead to a wide variety of responses. Bak envisaged that these marginally stable states are characterised by the lack of any typical time or length. This configuration is of a similar type to that seen in the thermodynamics of phase changes. The lack of typical scale leads to algebraic correlation functions associated with power-laws.[4] Hence, the distribution functions describing the frequency with which SOC events occur exhibit power-law characteristics. For our earthquake example, if E is the energy released during an earthquake, then the probability of an earthquake of that size is given by the power-law relationship $P(E) \sim E^{-B}$. As noted in [8], such a simple law should have an elegant explanation. We shall see later, in Chapter 3 when examining the evidence for such emergent behaviour in warfare, that the number of casualties in war also has such a power-law distribution. However, as yet no such elegant explanation is available, either for earthquakes or wars.

Clustering in Space and Time

The formation of clusters that are *fractal* in both space and time is common in natural systems, as we have already seen. It was this type of behaviour that Bak wanted to explain. We will see that these relate to ideas of *correlation* in space or time (in contrast to coincidence in space or time). Correlation in space or time is a signal of local clustering and collaboration spatially (e.g., across a battlespace) or in time (e.g., across an information grid–reading e-mail creates a correlation in time between individuals, taking a phone call creates a coincidence in time). The properties of fractals and their link to chaotic behaviour have been examined intensively over the last two decades.[5]

[4]A power-law $f(x) = x^a$ has the property that the relative change

$$\frac{f(kx)}{f(x)} = k^a$$

is independent of x. Power-laws, in this sense lack characteristic scale.

[5]See for example the standard text *Chaos and Fractals* by Heinz-Otto Peitgen, Hartmut Jürgens and Dietmar Sauape. Springer. 1992.

Despite this, very little (until now) has been known about why fractals form. Fractal structures are not the lowest energy configuration that can be selected in, for example, thermodynamic systems, therefore some kind of dynamic selection of configuration must be taking place.

Bak explains the connection in the following way. A signal will be able to evolve through the system as long as it is able to find a connected path of above-threshold regions. When the system is either driven at random or started from a random initial state, regions that are able to transmit a signal will form some kind of random network. This network is correlated by the interaction of the internal dynamics with the external field. The complicated interrelation between the two driving dynamics means that a complex, finely-balanced system is produced. As the system is driven, after this marginally stable self-organised state has been reached, we will see flashes of activity as external perturbations interact with internal drivers to spark off avalanches (i.e., clusters) of activity through different routes in the system. Bak's assertion is that the structure of this dynamic network is fractal. If the activated clusters consist of fractals of different sizes, then the duration of the induced processes travelling through these fractals will vary greatly. Different timescales of this type lead to what is termed *1/f noise.*

1/f noise is a label used to describe a particular form of *time correlation* in nature. If a time signal fluctuates in a seemingly erratic way, the question is whether the value of the signal $N(\tau_0)$ at time τ_0 has any correlation to the signal measured at time $\tau_0+\tau$ $(N(\tau_0+\tau))$. The amount of causation is characterised by a temporal correlation function:

$$G(\tau) = \langle N(\tau_0)N(\tau_0+\tau)\rangle_{\tau_0} - \langle N(\tau_0)\rangle_{\tau_0}^2 .$$

The correlation function, for a stationary process, is linked to the power spectrum as follows:

$$S(f) = 2\int_0^\infty d\tau G(\tau)\cos(2\pi f\tau)$$

where the power spectrum is defined in terms of the square amplitude of the Fourier transform of the time signal:

$$S(f) = \lim_{T\to\infty} \frac{1}{T}\left|\int_{-T}^{T} d\tau N(\tau) e^{2i\pi f\tau}\right|^2 .$$

1/f noise corresponds to the case where $S(f) = \frac{1}{f}$ and corresponds to a fractal clustering of the signal amplitude in time.

Fractal Clustering in Space

For a system that is not poised in a critical state and thus not about to change its mode of behaviour, the reaction of the system is described by a characteristic response time and characteristic length of scale over which the perturbation is felt. However, for a critical system, the same perturbation applied at different positions or the same position at different times can lead to a response of any size. The average is not therefore a useful measure of response. The amount of the system involved is a cluster in the spatial dimensions of the system.

An Example–A Pile of Dry Sand on a Beach

The nature of this critical state can be illustrated using a simple example–a pile of sand. Anyone can try this experiment on a sunny day by the seaside. Normally, the grains of sand only interact locally and nothing surprising happens. However, if the critical state (defined by the slope of the pile) is perturbed by the addition of one grain of sand to a randomly chosen position on top of the pile, the extra grain will induce an avalanche (a cluster) characterised by spatial and temporal characteristics such as the total number of sand grains s involved in the avalanche and the lifetime of the avalanche t. The statistical distributions describing the response are denoted by $P(s)$ and $P(t)$. In the critical state, we expect broad power-law distributions of the form $P(s) \sim s^{-\beta}$ and $P(t) \sim t^{-\alpha}$. The particular values of α and β are then characteristic of how the system can create such correlations in time and space. These distributions will typically be bounded by *spatial and temporal lower and upper cut-offs*. For example, an avalanche cannot involve the displacement of less than one grain of sand, and the duration of an avalanche cannot be shorter than the time it takes one grain to move a distance equal to the size of a single grain.

In the thermodynamic theory of phenomena, the spatial correlation function plays a fundamental role. If the system is described by a space-time field $n(\boldsymbol{r},t)$, the spatial correlation function is defined as:

$$G(r) = \langle n(\boldsymbol{r}_0)n(\boldsymbol{r}_0 + \boldsymbol{r})\rangle_{\boldsymbol{r}_0} - \langle n(\boldsymbol{r}_0)\rangle_{\boldsymbol{r}_0}^2$$

where a thermal average and an average over the position $\boldsymbol{r}_0$ is assumed.

Away from the critical temperature, the correlations decay exponentially, $G(r) \sim \exp(-r/\xi)$, beyond the correlation length ξ. The correlation length diverges, $\xi \sim |T - T_c|^{-\nu}$, as the critical temperature is approached. At the critical point, $T = T_c$, the correlation function changes functional behaviour from exponential to algebraic dependence upon $\boldsymbol{r}$, i.e.

$G(r) \sim r^{-\eta}$. The divergence is considered to be the signal of the lack of a characteristic length scale. Self-organising systems may be characterised and examined in this way.

Cellular Automata

The introduction in 1987 of the Self-Organised Criticality concept employed the language of avalanches (clusters) of sand grains. It proposed a simulation model (a *cellular automata* model to be precise) of the most essential features of sand dynamics. This cellular automata model is indeed characterised by power-laws and exhibits critical behaviour. As an example, let us look again at the modelling of an ecosystem consisting of a number of coevolving species (the Bak-Sneppen evolution model) that we mentioned earlier. This automaton process considers the points of a grid and has a simple set of rules determining how the system changes from one time-step to the next, to represent the changing fitness of the species itself, and the coevolutionary impact of that change on the fitness of closely linked species in the ecosystem. Note that this linkage is one of local species influence and coevolution. It does not necessarily assume physical closeness. Although these rules are simple, the emergent behaviour of the system is complex and surprising–a characteristic of such nonlinear interactions.

An Example: Clustering and Coevolution on a Global Information Grid

We can use the evolution model just described to gain insight into the effect of a Global Information Grid. Imagine such a grid in two dimensions. At each grid point is positioned an element of our force. Each such force element has a "fitness" value corresponding to its ability to evolve and adapt to local circumstances as a function of the information available on the grid. We assume these fitness values are random at first. At each step of the process, we assume that the force element with the smallest fitness is likely to have to adapt fastest to its local environment. In so doing, it will change the fitness values of the units closest to it on the information grid (i.e., there is local coevolution). Note that these force elements may be separated by large and varying distances in space. With these assumptions, over time the force elements will form clusters of coevolution of the form predicted by the Bak-Sneppen model. In particular, the statistics of emergent cluster size can be predicted mathematically to converge to a power-law. Thus, the probability of a cluster of size S, $P(S)$, is of the form $S^{-\tau}$, where τ is a characteristic cluster exponent for this particular information grid.

Thus far, we have looked at complex effects and correlations across space and time. What happens if we restrict ourselves to looking at the boundary between two different regimes (such as two different nationalities or two opposing armed forces), and how this would move over time depending on the local coevolution of the elements involved?

Movement of a Boundary

In natural systems, we can consider the movement of a boundary through a medium (for example, the boundary of an atomic surface, the boundary of a growing cluster of bacteria, or the front of advance of a fluid "invasion" of a medium such as a crystalline rock). This has been studied extensively in relation to the laying down of single atom surfaces using molecular beam epitaxy [9]. The most relevant case from our point of view is the front of advance of fluid "invasion" of a medium. As described in [9], we can represent the medium itself as consisting of a lattice of cells, each with either a 1 or 0 in it. A "1" represents the fact that that cell can be wetted. The proportion of cells containing a "1" is defined as p. For large configurations, we can also interpret p as the probability that a particular cell contains a "1." A "0" represents the fact that the cell cannot be wetted–it thus "pins" the advance of the fluid through the medium, at least locally. In the standard Direction Percolation Depinning (DPD) model, we start off with (for example) a two-dimensional square lattice of cells, some labelled with 1s and the rest with 0s to represent the distribution of these "pinning" forces throughout the medium. Initially, we wet one edge of the lattice. In Figure 1.5, we show the wetted right-hand edge. In the standard DPD model, one of the unpinned cells in the next column is chosen at random, and the fluid invades that cell.

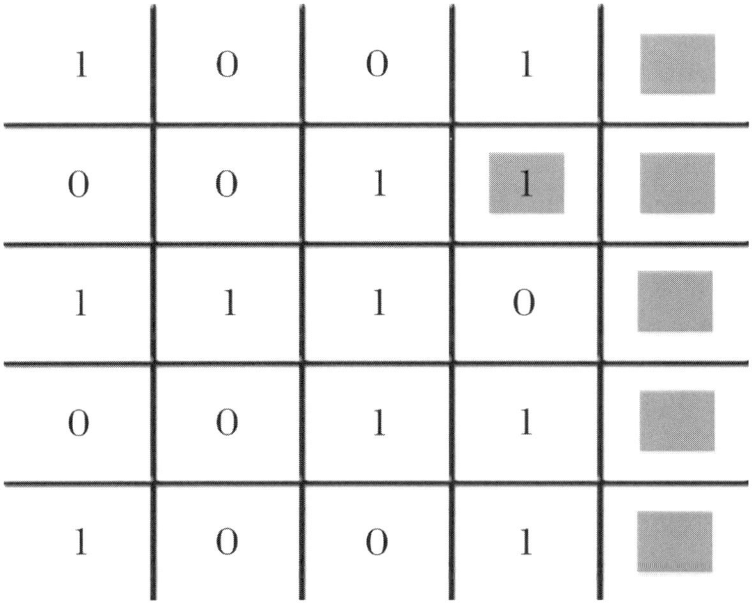

Figure 1.5: Local Pinning of a Fluid Boundary

It turns out that for this case, when the pinning probability p is greater than a critical value p_c, the growth of the interface is halted by a spanning path of pinning cells. Such models of interface or boundary movement exhibit fractal properties of the interface, as discussed in detail in [9]. We shall see similar effects later in our discussion in Chapter 4 of the control of the battlespace using ideas based on preventing the flow of opposing forces and/or third parties through the space. Rather than choosing the next cell to invade at random, as in the DPD model, we can use a model of the process that is more akin to the manoeuverist principle of applying your strength where the opponent is weak–in other words, the cell next to be wetted is the one where the local pinning force of the medium is weakest. Such a model of the boundary movement is the *Invasion Percolation model.* We can create a model of this process in a way that is consistent with our description of the Bak-Sneppen evolution model of local coevolution [6]. We start by assigning, as with the Bak-Sneppen evolution model, random numbers f_i between 0 and 1 to the points of a d-dimensional lattice. Initially, one side of the lattice is the wetted cluster. The random numbers at the boundary of the wetted cluster are examined. At each update step s, the site with the smallest random number $f_{i,s}$ on the boundary of the wetted area is located and added to the cluster. In this case, we can interpret the values f_i as the values of the local pinning force, and the cluster advances at those points where the pinning force is smallest. As noted in [6], an important physical realisation of invasion percolation is the displacement of one fluid by another in a porous medium. The boundary of the cluster created by this process is fractal and has a fractal dimension in the range 1.33-1.89 dependent on the exact definition of the boundary [6, Appendix].

Movement of a Boundary

AN ANALYTICAL MODEL OF THIS PROCESS

In natural systems, the boundary of such an interface that is moving through a medium can be characterised by its "roughness." This is defined as follows.

We assume that the interface is defined across a linear space of length L (as shown in Figure 1.6). The *roughness* of the interface at a given time t and for a span of length L, is defined [9] as:

$$W(L,t) = \sqrt{\frac{1}{L}\sum_{i=1}^{L}(h(i,t) - \bar{h}(t))^2}$$

where the interface length is divided into a number of natural cells of unit length, $h(i,t)$ is the height of the ith column at time t (as indicated in Figure 1.6) and $\bar{h}(t)$ is the average of these heights at time t. The roughness $W(L,t)$ is thus the standard deviation of the height at a given time. For many natural systems, the roughness first goes through a transition period before stabilising at an equilibrium value. Once it has stabilised, the expected behaviour is that this "saturation" value W_{sat} scales with the interface width L, i.e. we expect the relationship $W_{sat} \sqcup L^{\alpha}$.

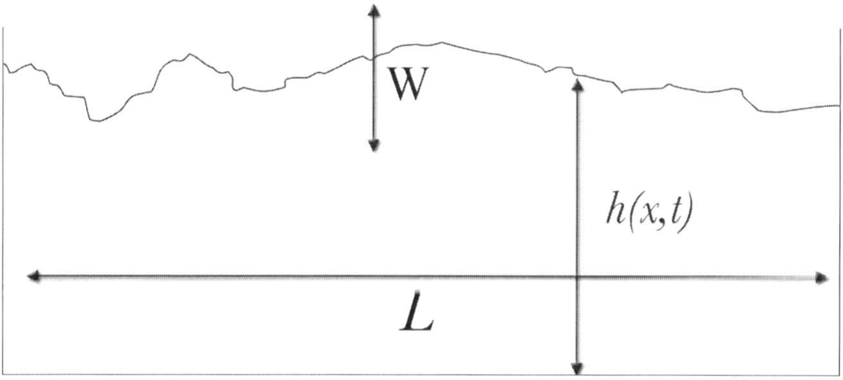

Figure 1.6: The Roughness "W" of an Interface

When this occurs, the exponent α is referred to as the *roughness exponent.* It is then a characteristic exponent of the invasion process under study. A typical value of α is about 0.6 (as shown in [9]). This idea can also be used to characterise Brownian motion, as we discuss later in this chapter in the context of stock price dynamics.

By using symmetry arguments, we can derive [9] an analytical expression of the rate at which such an interface moves through a medium. (In terms of conflict, this corresponds to the rate of advance of a combat front.) This is given by the Kardar-Parisi-Zhang (KPZ) equation:

$$\frac{\partial h(x,t)}{\partial t} = \nu\nabla^2 h + \frac{\lambda}{2}(\nabla h)^2 + \eta(x,t).$$

The first term in this equation represents linear effects of the interface growth, the second captures nonlinear effects, and the third is a noise term. This thus represents the starting point for an analytical expression (i.e., a metamodel) of the advance of a conflict front through a locally controlled area, as discussed below.

An Example: Force Control along a Boundary

The concept of pinning a fluid locally is similar to the idea of trying to exert local control over a boundary to prevent the flow of other forces or third parties across that boundary. We will see later (in Chapter 4) that the idea of control as the prevention of such flows through an area has important implications for the emergent behaviour of a force (or two competing forces) attempting to exert control over a battlespace.

Movement of a Boundary

In the case of control along a boundary, Complexity Theory, in terms of the invasion percolation model, can be used to analyse the effect of two forces (an attack and a defence force) interacting across a boundary, when the boundary moves at the point where the defending (pinning) force is weakest. If the defence pinning force is coevolving locally, then the boundary should form a fractal with a fractal dimension in the range 1.33-1.89 [6, Appendix], as we have seen. In Chapter 3, we show that there is historical evidence in warfare for such an effect, and for values in this range.

An Example of Complex Behaviour and Fractal Time Series in Economics

Brownian Motion, Fractals, and Similarity

Mandelbrot [10] has considered the movement of stock prices using fractal ideas and we start by using a simple example based on Brownian motion (since this is the basis of most current predictors of stock price volatility). Consider then Brownian motion in one space variable, thus the motion of particles is restricted to a line. The impacts affect the particle only from the left and right, causing a displacement of length l in either direction. Can any prediction be made about the total displacement after a number of time-steps n? First, the total expected displacement is zero, as all displacements are $+l$ or $-l$, both with equal probability 0.5. Consider, instead, the square of the displacement. The average of these square displacements, called the *mean square displacement*, indicates how much the particles spread in a given number of time-steps on the average. Its value is nl^2. In terms of Brownian motion, the number of steps corresponds to the number of impacts on a particle and cannot be directly measured in an experiment.

Consider, therefore, time duration t. Assuming an average number of n impulses during a timespan t, the particle travels a total length nl. If v is the average speed of the particle, then we have $vt = nl$. The mean square displacement nl^2 thus equals vlt. Therefore, the mean square displacement is proportional to the timespan. Is there anything else that can be said regarding the distribution of the displacement after time t? Experimentally, it is found that the statistics for a simulated one-dimensional Brownian motion follow a Gaussian or normal distribution. This is not surprising as a normal distribution arises where independent and identically distributed random events are averaged and is an implication of the Central Limit Theorem of Statistics.

Brownian motion is also associated with *scaling relationships* and *fractal behaviour.* As a nonfractal object is magnified, no new features are revealed. As a fractal object is magnified, finer details are revealed. The size of the smallest feature of a nonfractal object is called the *characteristic scale.* A measurement made at finer resolution will include more of these smaller pieces. Thus the value measured of a property will depend upon the resolution used to make the measurement. How a measured property depends on the resolution used to make the measurement is called the *scaling relationship.* A fractal object has features over a broad range of sizes. Fractal phenomenological characteristics are:

1. SELF-SIMILARITY: behavioural characteristics are "similar" at different resolutions.
2. SCALING: the value measured for a property depends upon the resolution at which it is measured.
3. DIMENSION: the dimension of an object gives a quantitative measure of self-similarity and scaling. It tells us

how many new pieces of an object are revealed as it is viewed at higher magnification.

4. NONSTATISTICAL PROPERTIES may be observed. Moments may be zero or nonfinite (e.g., the mean tends towards zero and variance tends towards infinity).

There are two types of self-similarity:

1. GEOMETRICAL: pieces of the object are exact smaller copies of the whole object.
2. STATISTICAL: the value of the statistical property $Q(r)$ measured at resolution r is proportional to the value of $Q(ar)$ measured at a resolution ar such that $Q(ar) = kQ(r)$. For statistical probability distribution functions (pdf), this implies that: pdf[$Q(ar)$] = pdf[$kQ(r)$].

Statistical self-similarity is possible in both space (spatial) and time (temporal).

In summary then, such self-similarity implies a scaling relationship. The simplest form of the scaling relationship is that the measured value of a property $Q(r)$ depends on the resolution used to make the measurement, as in the equation $Q(r) = Br^b$, i.e. $LogQ(r)=LogB+bLog\ r$, where B and b are constants. Hence for self-similarity, we observe log-log linear behaviour. Experimentally, this implies that for Brownian motion, there should be a scaling factor, r, that yields curves that are visibly identical, i.e. when scaling Brownian motion in time, by a factor say of t and in amplitude by a factor of r, we should see no difference. This transformation is called a *scaling collapse* since it has the effect of collapsing the curves at different scales of measurement on top of each other into one normalised relationship.

The scaling relationship in time may be determined theoretically for Brownian motion. It follows from the mean squared analysis that the mean squared displacement, Δ^2, of the Brownian motion *X(t)*, is described by $\Delta^2 \propto t$. Now consider the rescaled random function:

$$Y(t) = rX\left(\frac{t}{a}\right),$$

i.e. the graph of *X* is stretched in the time direction by a factor *a* and in the amplitude by *r*. The displacements in *Y* for time differences *t* are the same as those in *X* multiplied by *r* for corresponding time differences t/a. Thus, the squared displacements are proportional to r^2t/a. In order to ensure the same constant of proportionality as the original Brownian motion, we require $r^2/a = 1$ or $r = \sqrt{a}$. For example, when replacing *t* by $t/2$, i.e. stretching the graph by a factor of 2, we have a=2 i.e., $r = \sqrt{2}$.

In this broader context of fractal processes, ordinary Brownian motion is a random process *X(t)* with Gaussian increments and $\mathrm{var}(X(t_2) - X(t_1)) \propto |t_2 - t_1|^{2H}$ where *H* (the Hurst exponent) = ½. We can consider (as discussed in [9]) the Brownian motion time series as describing an interface (between the parts above and below the series) stretching between the time points t_1 and t_2. In terms of our previous discussion of interface roughness, we now have $L = t_2 - t_1$. The standard deviation (i.e., the roughness of the interface generated by the Brownian motion time series) over this timespan *L* has the form L^α where in this case α equals the Hurst exponent *H* (as we can see from the expression for the variance above in terms of *H*). Thus, standard Brownian motion corresponds to a roughness exponent of ½. Other values of *H* are possible, corresponding to rougher or smoother forms of time series.

Stock Prices and Brownian Motion

Mandelbrot considered the movement of stock prices using a Brownian random walk process as his starting point [10]. The behaviour of such a variable z can be understood by considering the changes in its value in small intervals of time. Consider such a small interval of time Δt and define Δz as the change in z during Δt. Δz must have two basic properties: first, $\Delta z = \varepsilon\sqrt{\Delta t}$ where ε is a random draw from a standardised normal distribution (i.e., a normal distribution with mean zero and standard deviation of 1.0); and second, the values of Δz for any two different short intervals of time Δt are independent. It follows that Δz has a normal distribution with mean zero, standard deviation of $\sqrt{\Delta t}$ and variance Δt. The second property implies that z follows a Markov process. Now consider the change in the value of z during a relatively long period of time T. This can be denoted by $z(T)\text{-}z(0)$. This change can be regarded as the sum of the changes in z in N small time intervals of length Δt, where $N = \frac{T}{\Delta t}$. Thus:

$$z(T)\text{-}z(0) = \sum_{i=1}^{N} \varepsilon_i \sqrt{\Delta t}$$

where ε_i ($i = 1,2,...N$) are random drawings from a standardised normal distribution. It follows that $z(T)\text{-}z(0)$ is normally distributed with mean zero, variance of $N\ \Delta t = T$ and standard deviation $\sqrt{T}$. The process described so far has a drift rate of zero and a variance rate of 1. The expected value of z at any future time is equal to its current value and the variance of the change in z in a time interval T equals T.

A *generalised Wiener process* for a variable x generalises the concept of the drift rate and variance of such a process, and may be defined in terms of dz as $dx = adt + bdz$ where a and b are constants. The *adt* term implies that *x* has an expected drift rate of *a* per unit time. Without the *bdz* term, the equation becomes:

$$dx = adt \text{, i.e. } \frac{dx}{dt} = a .$$

The *bdz* term may be regarded as adding noise or variability to the path followed by *x*. For a small time interval, Δt, the change in *x*, Δx, is given by $\Delta x = a\Delta t + b\varepsilon\sqrt{\Delta t}$ where ε is a random draw from a standardised normal distribution. Δx has a normal distribution with mean of $a\Delta t$, standard deviation of $b\sqrt{\Delta t}$, and a variance of $b^2\Delta t$. Similarly, the mean change in *x* for any time interval *T* is normally distributed with mean change in *x* given by *aT*, standard deviation of change in *x* given by $b\sqrt{T}$, and variance of change in *x* as b^2T.

Even more generally, a stochastic process is defined as an *Ito process* if $dx = a(x,t)dt + b(x,t)dz$.

A basic simulation of stock price movement would be to use a generalised Wiener process. This is clearly inadequate as this assumes that both a constant drift rate and constant variance rate occur, i.e. the percentage stock return is dependent upon stock price. The constant expected drift rate assumption is inappropriate and is replaced by the assumption that the expected drift, expressed as a proportion of the stock price, is constant. Thus, if *S* is the stock price, the expected drift rate in *S* is μS for some constant parameter μ and for a small time interval, Δt, the expected change in *S* is $\mu S\Delta t$. If the variance rate of the stock price is always zero, then:

$$dS = \mu Sdt \quad \text{or} \quad \frac{dS}{S} = \mu dt$$

$$\text{i.e., } S = S_0 e^{\mu t}$$

where S_0 is the stock price at time zero. This indicates that when the variance rate is zero, the stock price grows (or declines) at a continuously compounded rate of μ per unit time. In practice, as stock prices exhibit volatility, a reasonable assumption is that the variance of the percentage return in a short period of time, Δt, is the same regardless of stock price. Define σ^2 as the variance rate of the proportional change in the stock price. Thus $\sigma^2 \Delta t$ is the variance of the proportional change in stock price, *S*, during time Δt. The instantaneous variance rate for *S* is $\sigma^2 S^2$. This implies that *S* can be represented by an Ito process that has an instantaneous drift rate μS and instantaneous variance rate $\sigma^2 S^2$. This can be written:

$$\mathrm{dS} = \mu \mathrm{Sdt} + \sigma \mathrm{Sdz}$$

or

$$\frac{\mathrm{dS}}{\mathrm{S}} = \mu\, dt + \sigma\, dz$$

and

$$d \ln S = \left(\mu - \frac{\sigma^2}{2} \right) dt + \sigma dz .$$

The change in ln *S* between times *t* and *T* is thus normally distributed:

$$\ln S_T - \ln S \approx \phi \left[\left(\mu - \frac{\sigma^2}{2} \right)(T-t), \sigma \sqrt{T-t} \right]$$

where S_T is the stock price at a future time *T*, *S* is the stock price at the current time, and $\phi(m,s)$ denotes a normal distribution with mean *m* and standard deviation *s*.

It follows that:

$$\ln S_T \approx \phi\left[\ln S + \left(\mu - \frac{\sigma^2}{2}\right)(T-t), \sigma\sqrt{T-t}\right]$$

thus ln S_t is normally distributed so that S_T has a log-normal distribution.

Under closer examination, stock prices in fact depart from log-normal behaviour. Examination of movements in stock prices presents changes greater than our model predicts. Stock returns exhibit *leptokurtosis*, i.e. the likelihood of returns near the mean and of large returns is greater than our Brownian motion model predicts, whilst other returns tend to be less likely [11, 12]. Table 1.1 gives a précis of the statistics of this non-normal behaviour.

Decade	Mean %	Standard Deviation %	Skewness	Kurtosis	Max %	Min %
1920s	0.0058	2.1453	0.1321	11.9272	11.6396	-13.7203
1930s	-0.0216	1.8233	0.2856	4.5422	13.8635	-8.7776
1940s	0.0098	0.7618	-1.0890	12.8516	6.5275	-7.0431
1950s	0.0470	0.6578	-0.9201	7.3343	4.0476	-6.7660
1960s	0.0063	0.6558	0.0437	5.4873	4.5787	-5.8815
1970s	0.0025	0.9262	0.2693	1.7889	4.9517	-3.5660
1980s	0.0489	1.1607	-4.3115	99.3933	9.6661	-25.6315
Overall	0.0142	1.1329	-0.7919	28.8256	13.8635	-25.6315

Table 1.1: Decade-by-Decade Behaviour of Daily Returns from the Dow Jones Index

Clustering in Time

What is of particular interest is that Turner and Weigel's data strongly suggest the occurrence of temporal clustering. Over the 1928-1989 period, 12.5 and 37.5 percent of all extreme positive jumps in the S&P 500 occurred within one and five days respectively of another positive jump in equity prices. Positive jumps in the Dow Jones were similarly clustered with 11.3 percent of the positive jumps taking place within one day, and 36.2 percent transpiring within 5 days of each other.

The second defect with the Brownian motion model is that if the model holds, then stock returns should be proportional to elapsed time and the standard deviation of returns should be proportional to the square root of elapsed time. This is based upon the scaling properties of Brownian motion. Turner and Weigel's data also demonstrated that monthly and quarterly volatilities are higher than annual volatilities and conversely, that daily volatilities are lower than annual volatilities, i.e. their research shows that the stock returns do not scale in a Brownian motion sense.

The answer to these problems can perhaps be found by considering the mathematical form of these distributions [13]. We have shown that scaling is observed within Brownian motion and that this may be characterised by a Gaussian probability distribution plot. A log-normal distribution may be crudely assumed to be a Gaussian plot with an increased tail. The tails of Pareto distributions also die off much more slowly than Gaussian or log-normal tails. Such *fat tail* probability distributions are thus increasingly describing the greater and greater volatility of the system. Current financial models rely on an extension of the basic Brownian motion model, either assuming a different volatility distribution (as in "jump volatility calculations") or attempting to empirically analyse and fit to the observed volatility. For example, for a Pareto probability distribution of the random variable y, with $y>1$ then we obtain:

$$F(y) \sqcup \frac{1}{y^{\alpha}},$$

i.e. a power-law. For further discussion of these and related ideas from Complexity Theory and dynamical systems used in financial mathematics, see references 10 and 13.

Summary

In summary, we have looked in some depth at the complex behaviour of natural biological and physical systems. From our analysis of these open and dissipative systems, it is clear that there are a number of key properties of complexity that are important to our consideration of the nature of future warfare. Such futures, involving the exploitation of loosely coupled command systems such as Network Centric Warfare, will have to take account of these key properties. A list of these is given here, and then discussed further in Chapter 2 in the context of Network Centric Warfare.

1. Nonlinear Interaction: this can give rise to surprising and non-intuitive behaviour, on the basis of simple local coevolution.
2. Decentralised Control: the natural systems we have considered, such as the coevolution of an ecosystem or the movement of a fluid front through a crystalline structure, are not controlled centrally. The emergent behaviour is generated through local coevolution.
3. Self-Organisation: we have seen how such natural systems can evolve over time to an attractor corresponding to a special state of the system, without the need for guidance from outside the system.

4. NONEQUILIBRIUM ORDER: the order (for example, the space and time correlations) inherent in an open, dissipative system far from equilibrium.

5. ADAPTATION: we have seen how such systems are constantly adapting–clusters or avalanches of local interaction are constantly being created and dissolved across the system. These correspond to correlation effects in space and time, rather a top-down imposition of large-scale coincidences in space and time.

6. COLLECTIVIST DYNAMICS: the ability of elements to locally influence each other, and for these effects to ripple through the system, allows continual feedback between the evolving states of the elements of the system.

REFERENCES

1 HOFFMAN F G and HORNE G E (1998). *Maneuver Warfare Science 1998.* Dept of the Navy, HQ U.S. Marine Corps, Washington DC.

2 FORDER R (2000). "The Future of Defence Analysis." *Journal of Defence Science.* 5, No. 2. pp. 215-226.

3 CEBROWSKI A (2000). "Network Centric Warfare and Information Superiority" Keynote address from proceedings, Royal United Services Institute (RUSI) conference "C4ISTAR; Achieving Information Superiority." July 2000. RUSI, Whitehall, London, UK.

4 PRIGOGINE I (1980). *From Being to Becoming; Time and Complexity in the Physical Sciences.* W H Freeman and Co., San Francisco, USA.

5 MOFFAT J (2002). *Command and Control in the Information Age; Representing its Impact.* The Stationery Office, London, UK.

6 PACZUSKI M, MASLOV S, and BAK P (1996). "Avalanche Dynamics in Evolution, Growth and Depinning Models." *Physics Review.* E 53 No. 1. pp. 414-443.

7 BAK P, TANG C, and WIESENFELD K (1987). *Self-Organised Criticality; An Explanation for 1/f Noise.* Physics. Review Letters, 59. pp. 381-384.

8 SETHNA J P, DAHMEN K A, and MYERS C R (2001). "Crackling Noise." *Nature.* Vol 410. pp. 242-250.

9 BARABASI A L and STANLEY H E (1995). *Fractal Concepts in Surface Growth.* Cambridge University Press. Cambridge, UK.

10 MANDELBROT B (1997). *Fractals and Scaling in Finance.* Springer-Verlag.

11 TURNER A L and WEIGEL E J K (1990). "An Analysis of Stock Market Volatility." Technical Report, Frank Russell Co., Tacoma WA. USA.

12 TURNER A L and WEIGEL E J K (1992). "Daily Stock Market Volatility: 1928-1989." *Management Science.* 38. pp. 1586-1609.

13 MANTEGNA R and STANLEY H E (2000). *An Introduction to Econophysics; Correlations and Complexity in Finance.* Cambridge University Press. Cambridge, UK.

ADDITIONAL REFERENCE

14 PEITGEN H-O, JURGENS H, and SAUAPE D (1992). *Chaos and Fractals.* Springer-Verlag.

CHAPTER 2

CONCEPTS FOR WARFARE FROM COMPLEXITY THEORY

As a starting point for our journey, Chapter 1 established in some depth and detail the key ideas and methods from Complexity Theory that we can bring to bear in thinking about and modelling future warfare. In his RUSI keynote talk [1], ADM Cebrowski (Head of the Office of Force Transformation, U.S. DoD) indicated that Network Centric Warfare is an emerging theory of war based on the concepts of nonlinearity, complexity, and chaos. It is less deterministic and more emergent; it has less focus on the physical than the behavioural; and it has less focus on things than on relationships. It is clear from the discussion and modelling in Chapter 1 that Complexity Theory is the essence of these ideas.

In a previous book [2] we showed how it is possible to capture the effects of command and control in agent-based simulations of Information Age warfare. This is done by representing the process as the interaction of top-down and bottom-up effects. These are described as Deliberate Planning and Rapid Planning. Deliberate Planning is appropriate when ample time is available for the consideration of a number of alternative courses of action by either "side" and a course of action can be chosen that is considered to be, in some sense, optimal. Rapid Planning is appropriate when time is short and expert decisionmaking under stress leads to a pattern-matching approach. To quote from [2]:

> *Combat is, by its nature, a complex activity. Ashby's Law of Requisite Variety...which emerged from the theoretical consideration of general systems as part of Cybernetics, indicates that to properly control such a system, the variety of the controller (the number of accessible states which it can occupy) must match the variety of the combat system itself. The control system itself, in other words, has to be complex. Some previous attempts at representing C2 in combat models have taken the view that this must inevitably lead to extremely complex models. However, recent developments in Complexity Theory...indicate another way forward. The essential idea is that a number of interacting units, behaving under small numbers of simple rules or algorithms, can generate extremely complex behaviour, corresponding to an extremely large number of accessible states, or a high variety configuration, in Cybernetic terms. It follows that, if we choose these simple interactions carefully, the resultant representation of C2 will be sufficient to control, in an acceptable way, the underlying combat model. As part of this careful choice, we need to ensure that the potentially chaotic behaviour generated by the interaction*

of these simple rules is 'damped' by a top-down C2 structure which remains focused on the overall, high level, campaign objectives....

...It follows, from what we have just said, that the representation of the C2 process must reflect two different mechanisms. The first is the lower-level interaction of simple rules or algorithms, which generate the required system variety. The second is the need to damp these by a top-down C2 process focused on campaign objectives. Each of these has to be capable of being represented using the same Generic HQ/Command Agent object architecture. We have chosen to do this by following the general psychological structure of Rasmussen's Ladder, as a schema for the decisionmaking process. At the lower levels of command (below about Corps, and equivalent in other environments), this will consist of a stimulus/response mechanism. In cybernetic terms, this is feedback control. At the higher level, a broader (cognitive-based) review of the options available to change the current campaign plan (if necessary) will be carried out. In cybernetic terms, this is feedforward control since it involves the use of a 'model' (i.e., a model within our model) to predict the effects of a particular system change.

In the last chapter of [2] (Chapter 6: "Paths to the Future") the following point is made, which is the foundation for all of the work and ideas presented here:

Modelling and analysis to determine the effect of such phenomena underpin our thinking about such future conflict, the representation of information and command being at their heart. A new approach to capturing these effects has been put forward in this book, and is having a significant influence on the approach to

> *modelling these phenomena. However, capturing the process of intelligent agents in conflict, set within a widely divergent set of possible futures, leads to a rich set of possible trajectories of system evolution for analysis to consider. We thus need to complement this effort with other work to categorise and understand the classes of behaviours which might emerge from such a complex situation. This is the domain of Complexity Theory.*

The overall aim is thus to develop an "Operational Synthesis" (as discussed in [3]) of both agent-based modelling approaches (as described in [2]) and higher level mathematical metamodels based on Complexity Theory. Reference [2] lays out some initial ideas on how to develop such an understanding based on a theoretical approach to the development of a higher level "metamodel" of a cellular automaton model of conflict such as the ISAAC model developed by the U.S. Marine Corps Combat Development Centre. We will develop these ideas further in Chapters 4 and 5. In Chapter 5 in particular, we will consider again the ISAAC model and demonstrate how the ideas of Complexity Theory lead to an understanding of the clusters forming and reforming in such a model, and how they relate to the emergent behaviour of the model.

For the moment, let us consider the list of key concepts from Complexity Theory at the end of Chapter 1. These are on the left-hand side of Table 2.1. On the right-hand side is an interpretation in terms of military behaviour and doctrine (which we have termed "an Information Age force structure").

The nature of Network Centric Warfare for such future Information Age forces can be outlined as: within a broad intent and constraints available to all the forces, the local force units self-synchronise under mission command in order to achieve the overall intent [4].

COMPLEXITY CONCEPT	INFORMATION AGE FORCE
Nonlinear interaction	Combat forces composed of a large number of nonlinearly interacting parts.
Decentralised Control	There is no master "oracle" dictating the actions of each and every combatant.
Self-Organization	Local action, which often appears "chaotic," induces long-range order.
Nonequilibrium Order	Military conflicts, by their nature, proceed far from equilibrium. Correlation of local effects is key.
Adaptation	Combat forces must continually adapt and coevolve in a changing environment.
Collectivist Dynamics	There is a continual feedback between the behaviour of combatants and the command structure.

Table 2.1: Relation Between Complexity and Information Age Warfare

This process is enabled by the ability of the forces involved to robustly network. We can describe such a system as *loosely coupled* to capture the local freedom available to the units to prosecute their mission within an awareness of the overall intent and constraints imposed by high-level command. This also emphasises the looser correlation and *nonsynchronous* relationship between inputs to the system (e.g., sensor reports) and outputs from the system (e.g., orders). In this process, information is transformed into "shared awareness," which is available to all. This leads to units linking up with other units, which are either local in a physical sense or local through (for example) an information grid or Intranet (self-synchronisation). This in turn leads to emergent behaviour in the battlespace, as shown in Figure 2.1.

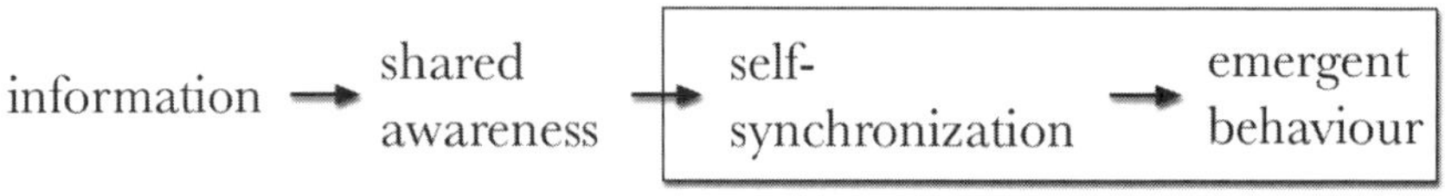

Figure 2.1: Information Leading to Emergent Behaviour

Compare these ideas with the broad conceptual framework of Complexity Theory, as summarised below.

The Conceptual Framework of Complexity

Prof. Murray Gell-Mann [5] traces the meaning to the root of the word. *Plexus* means braided or entwined, from which is derived *complexus*, meaning braided together, and the English word *complex* is derived from the Latin. Complexity is therefore associated with the intricate intertwining or inter-connectivity of elements within a system and between a system and its environment. In a human system, connectivity means that a decision or action by any individual (group, organisation, institution, or human system) will affect all other related individuals and systems. That effect will not have equal or uniform impact, and will vary with the state of each related individual and system at that time. The state of an individual and system will include its history and its constitution, which in turn will include its organisation and structure. Connectivity applies to the interrelatedness of individuals within a system, as well as to the relatedness between human social systems, which include systems of artifacts such as information systems and intellectual systems of ideas.

The term *complexity* is used to refer to the theories of complexity as applied to *complex adaptive systems* (CAS). These are dynamic systems able to adapt and change within, or as part of, a changing environment. It is important to note, however, that there is no dichotomy between a system and its environment in the sense that a system always adapts to a changing environment. The notion to be explored is rather that of a system closely linked with all other related systems making up an "ecosystem." Within such a context, change needs to be seen in terms of coevolution with all other related systems as we saw in Chapter 1, rather than as adaptation to a separate and distinct environment.

Collectivist Dynamics and Feedback

The phenomenological definition of a complex system is that it exhibits nonlinear, emergent, adaptive behaviour. Nonlinear behaviour is associated with far-from-equilibrium, open systems, in that cause and effect are no longer linearly connected. This is ultimately due to the type of internal-external system interactions (feedback) affecting our system.

Self-Organisation and Clustering

Self-organisation in this context is taken to mean the coming together of a group of individuals to perform a particular task. They are not directed by anyone outside the group. This is not the same as "self-management," as no manager outside the group dictates that those individuals should belong to that group, what they should do, or how it should be done. It is the group members themselves who choose to come together, who decide what they will do and how it will be done. A feature of these groups is that they are informal and often temporary. Enabling self-organisation can often be a source of innovation. Military commanders who understand the nature of *auftragstaktik*[1] have always understood this: a commander must regard his superior's intention as sacrosanct and make its attainment the underlying purpose of everything he does. He is given a task and resources and any constraints, and within this framework he is left to make his plan.

We already know from Chapter 1 that some complexity models of natural ecosystems use extremal dynamics as a behavioural driver. In systems where the dynamical evolution is a struggle against various types of thresholds or barriers, the action will predominately occur where the net barrier to change is the smallest. The Bak-Sneppen evolution model described in Chapter 1 is an example of such an extremal model. The species with the lowest fitness coevolves first. Similarly, in considering the movement of a fluid through a medium, the boundary moves where the pinning force is smallest. The avalanches of active sites are the clusters created in such models. In the model ecosystem, the system self-organises towards a critical point where it has the greatest dynamical freedom–clusters of all sizes can potentially be created, and the statistics of the emergent distribution of such cluster sizes can be predicted–the distribution is of power law form. Such extremal dynamics echo one of the key tenets of manoeuvre warfare–namely, to focus your strength against your enemy's weakness.

Forest Fires, Clusters of Trees, and Casualties in War

The "forest fire" model is another example of a system that evolves to a critical point through a process of local interaction. In this case, we start with an empty two-dimensional grid. At each iteration of the process, with a certain probability p

[1]Auftragstaktik is directed control, as opposed to befehlstaktik (detailed order tactics).

(normally close to one), we drop a tree onto a random grid point. If the grid point is blank, the tree is planted. If there is already a tree there, the new tree is discarded. With probability *(1-p)* we drop a spark onto a random site instead of a tree. If the site is bare the spark goes out. If there is a tree on the site, the tree and all of its immediate neighbours burn. All of the neighbours of these trees then burn and so on until the complete cluster of linked trees is burned (this is termed a *forest fire*). The rate (which is *[1-p]* if each iteration of the process is counted as a unit of time) of sparks dropping onto the grid is termed the *sparking frequency*. This sparking frequency *sp* is a key driver of the dynamics of the forest ecosystem. If *sp* is small, very large clusters of trees are allowed to form, which span the entire grid. When a spark is then dropped, the forest fire wipes out an entire forest stretching from one side of the grid to the other. In Complexity Theory, this is known as *snapping noise* [6]. This name comes from looking at the behaviour of the system over time–large spikes of tree extinction (forest fires) are created at isolated points in time. If the sparking frequency *sp* is very large, then tree clusters do not have the chance to grow. Thus, over time, the system produces a large number of small spikes of activity, which are called *popping noise*. When *sp* is in the intermediate regime, the system self-organises to a critical state where the clusters of burnt trees have a distribution represented by a power law, and clusters of all sizes can be created. Over time, the spikes produced by this process (i.e., the time evolution of forest fires of various sizes) have a similar dynamic to that produced by the acoustic dynamics of crumpling paper [6], and so this regime is termed *crackling noise*.

It is possible [7] to relate such self-organised behaviour of a forest fire model to the statistics of the scale and intensity of conflicts. This is the beginning of an explanation as to why casualties in war follow a power law distribution. As noted in [7], "...the behaviour of the forest fire model can be explained in terms of a cascade model. If trees are randomly planted on a grid, the distribution of cluster sizes is exponential (Poissonian), not power law (fractal). The distribution of cluster sizes in the forest fire model is power law (fractal). This is because clusters of trees continuously grow and combine to form larger clusters. Small fires sample this population of clusters, but the loss of trees in fires is dominated by the largest fires. There is a self-similar cascade of trees from small to large clusters. In terms of the forest fire model, a spark ignites a tree and the model fire consumes the entire cluster to which this tree belongs. This is also the case for real forest fires. Ignition of the forest must take place for a fire to take place, and the fire will then spread through the contiguous flammable material. A war must begin in a manner similar to the ignition of a forest. One country may invade another country, or a prominent politician may be assassinated. The war will then spread over the contiguous region of metastable countries...." We will look at this fractal and power law behaviour of casualties in more detail in Chapter 3.

Tuning and Goal Seeking

This leads us to consider the concept of *tuning*. Self-organising systems (such as the Bak-Sneppen evolution model of an ecosystem described in Chapter 1) can, as their name implies, develop local organisation within the system in order to evolve towards an attractor. Tuning can be seen as a directive way for the macrosystem to attempt to influence the behaviour of the microsystem. A controlling intelligence is deemed to be necessary in order to guide the system towards a particular goal. An example of this is the tuning of the sparking frequency model parameter *sp* in the forest fire model in order to obtain criticality (and hence power law statistics for the size of clusters of burnt trees). Varying the tuning parameter (the sparking frequency) of the forest fire model represents intervention from outside the system in order to ensure that it heads towards a particular goal. This question of tuning makes us consider the *boundaries* of the systems we are examining, and the flux of energy and/or information across the system boundary.

Information Flux across the Boundary of an Open System

In such open dissipative systems, there will always be fluxes of information and/or energy across the system boundary. As an example of this, we show in Chapter 4 how it is possible to quantify the flux of information across the boundary of an open system (a wargame) using the concept of Information Entropy. We can then use this to relate such a flux of information to emergent properties of the wargame (such as the number of casualties suffered). The idea of using entropy as a measure of information (and hence knowledge) is then applied in Chapter 4 to show how the benefits of network-centric (as opposed to platform-centric) approaches to specific task prosecution can be quantified (this latter based on work by the RAND Corporation).

REFERENCES

1 CEBROWSKI A (2000). "Network Centric Warfare and Information Superiority." Keynote address from proceedings, Royal United Services Institute (RUSI) conference. "C4ISTAR; Achieving Information Superiority." RUSI. Whitehall, London, UK.

2 MOFFAT J (2002). *Command and Control in the Information Age; Representing its Impact.* The Stationery Office. London, UK.

3 HOFFMAN F G and HORNE G E (1998). *Maneuver Warfare Science 1998.* Dept of the Navy, HQ U.S. Marine Corps. Washington, DC.

4 ALBERTS D S, GARSTKA J J and STEIN F P (1999). *Network Centric Warfare; Developing and Leveraging Information Superiority.* CCRP, DoD. Washington, DC, USA.

5 GELL-MANN M (1994). *The Quark and the Jaguar.* WH Freeman. New York, USA.

6 SETHNA J P, DAHMEN K A, and MYERS C R (2001). "Crackling Noise." *Nature.* 410 8. March 2001. pp. 242-250.

7 ROBERTS D C and TURCOTTE D L (1998). "Fractality and Self-Organised Criticality of Wars." *Fractals.* 6 No 4. pp. 351-357.

CHAPTER 3

EVIDENCE FOR COMPLEX EMERGENT BEHAVIOUR IN HISTORICAL DATA

INTRODUCTION

The first point to make is that the exploitation of manoeuvre warfare is not new. Commanders have exploited such an approach in previous generations. For example, LTG Sir Francis Tuker [1] indicated that at a three-dimensional spatial level, manoeuvre warfare is determined by three conditions:

1. Flanks shall be tactically open or it shall be possible to create a flank by break-in and breakthrough.
2. The mobile arm shall be predominant.

3. It shall be possible to administer the mobile arm to the point where it will decide the battle and gain decisive victory.

Examination of historical data should thus give us some insight into these effects.

Secondly, the idea that execution of military network-centric enterprises within a battlespace is of a self-organising nature has interesting consequences for the future of both warfighting and other operations. A system that comprises of clusters grouping and regrouping, the increasing tempos of battlespace awareness, operations, and responsiveness etc. require that the command decision process (or better, the rate of change of decision process) is undertaken with a similar tempo. The ratio of "battle speed" to "C2 speed"" is thus a critical issue, as we discussed in [2]. When this ratio is high, the system reverts to being self-organising in nature. In considering such self-organising systems earlier, we indicated that under certain conditions, the time series of outputs from such a system should have a certain fractal character. In the case of a model ecosystem such as the Bak-Sneppen evolution model discussed in Chapter 1, in fact, this fractal process is driven by the extremal (smallest) fitness values on the lattice of interacting species.

Time Series Behaviour

Applications of these ideas have been made in order to predict combat casualty rate patterns from WWII data [3]. A time series model based on this fractal self-organising approach has been contrasted with three other prediction methodologies: a neural network; use of nonlinear prediction (a prediction is made by searching the data for the nearest N points in a d-dimensional embedding and estimating the behaviour characteristics of the data for the next timestep [4]); and lastly, use of a maximum entropy method to calculate a power spectrum from which a linear prediction may be made [5]. All of these approaches are available in a package of time series analysis procedures (the Chaos Data Analyser) produced by the American Institute of Physics for the analysis of experimental data in natural systems, and that is what we have used here. A range of the data points from the original data were deleted (those at the end of the time series) and a prediction made of these data points, which is then compared with the original data for the 2nd Armoured Division. The plots (Figure 3.1 to 3.5) are of casualties per 1000 on the *y*-axis and time in days on the *x*-axis.

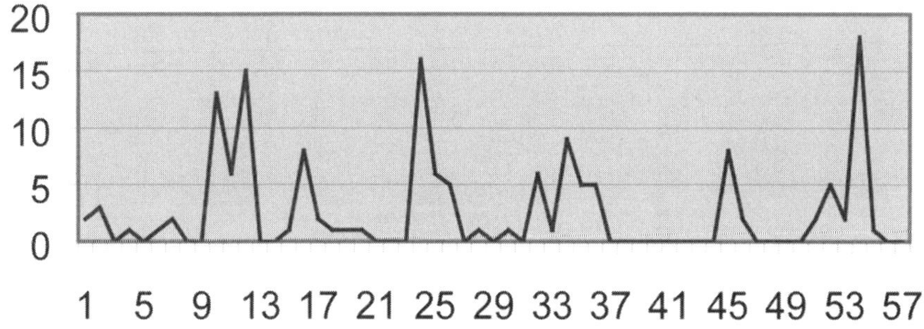

Figure 3.1: 2nd Armoured Division Data

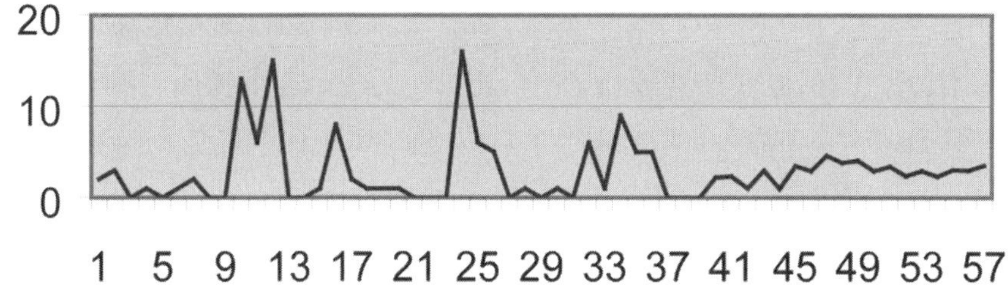

Figure 3.2: 2nd Armoured Division–Power Spectrum Prediction

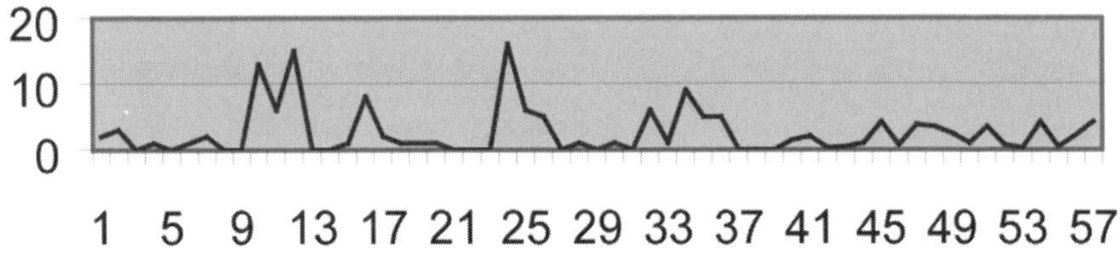

Figure 3.3: 2nd Armoured Division–Nonlinear Prediction

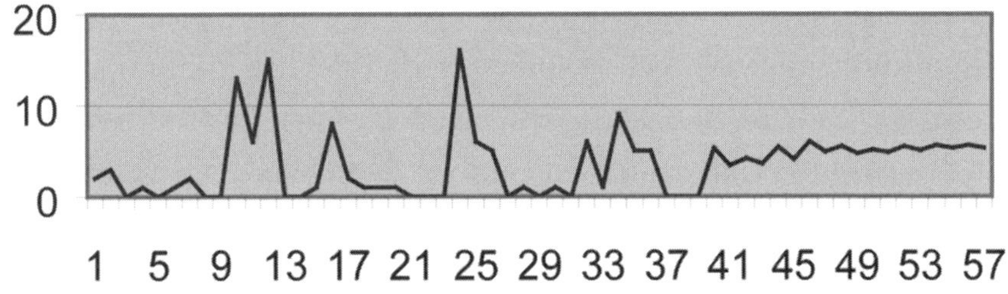

Figure 3.4: 2nd Armoured Division–Neural Net Prediction

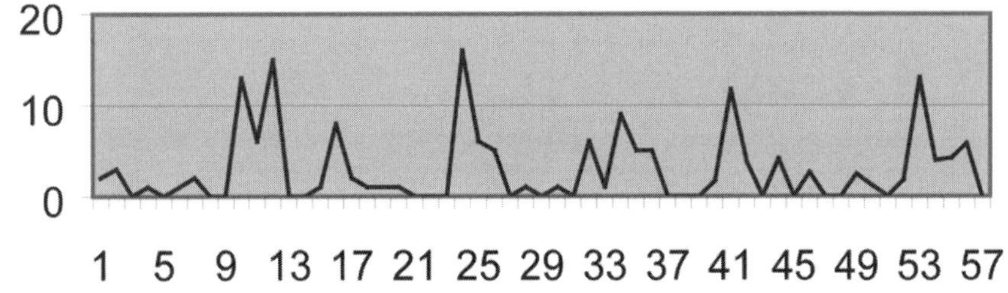

Figure 3.5: 2nd Armoured Division–SOC Prediction

The first part of this time series (up to day 38) was in fact used to train a number of different time series prediction methods, and these have been compared with the predictions for the days 39 onward. In fitting a prediction based on a self-organising criticality (SOC) fractal series, we have assumed that the circumstances remain sufficiently constant that we can fit a single SOC process (this corresponds to a power spectrum that is linear when plotted on a log-log scale). Comparing the "jerkiness" of the SOC prediction and the real data, the general pattern of the process is very similar.

We repeated the process with another data set drawn from Kuhn for the 9th Armoured Division, with similar results as summarised below.

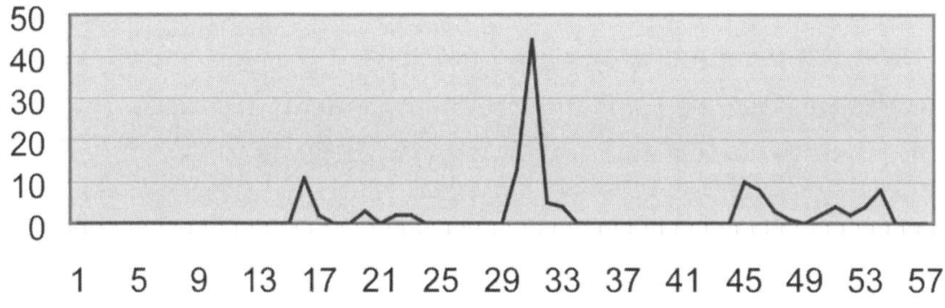

Figure 3.6: 9th Armoured Division Data

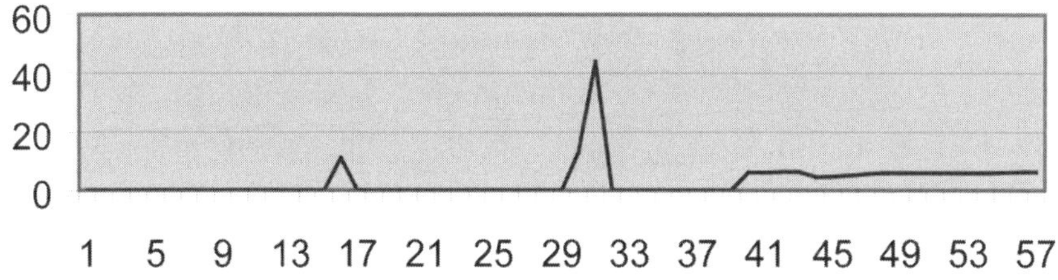

Figure 3.7: 9th Armoured Division–Neural Net Prediction

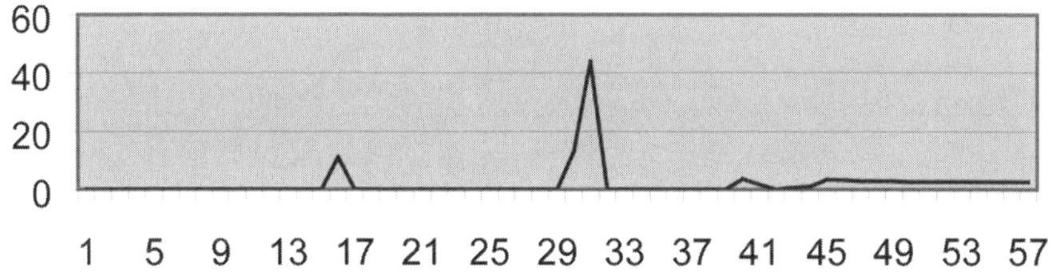

Figure 3.8: 9th Armoured Division–Power Spectrum Prediction

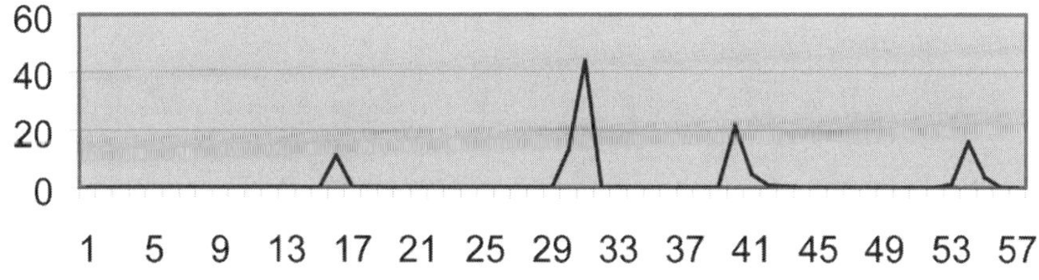

Figure 3.9: 9th Armoured Division–SOC Prediction

These plots indicate that the assumption of self-organisation appears to give casualty behaviour that is of the same form as for the real data, at least for these data points.

Kuhn himself suggested that the casualty rate data pattern, taken from his collection of combat rates from World War II to the present, displayed a move away from linearity. First, the number of "hotspots" in combat did not increase with larger force concentrations, and second, combat was characterised by high rates of casualties lasting for short periods of time, interspersed with low casualty rates. These results indicated, in turn, a move away from the type of modelling associated with attrition warfare. Quantitative patterns were also found to be based upon two types of operational forms: that of a *continual* front (characterised by combat on the Western Front in World War II) and that of a *disrupted* front, (characterised by combat on the Eastern Front in World War II). Combat casualties dramatically increased as a consequence of breakthrough operations in disruptive types of operations. Combat, therefore, could be seen as a process wherein quiet states are interspersed with "critical" irruptions.

Lauren [6] has also considered casualty data from World War II and compared it with outputs from the MANA model (an agent-based simulation similar in character to the ISAAC model we will discuss in detail in Chapters 4 and 5). He has shown that such data display fractal properties and power spectra that confirm the analysis of casualty data discussed in this chapter.

Further Historical Data on the Processes of "Irruption" and Breakthrough

In an historical analysis study [7] of the operational level of combat, it was found by Rowland that the occurrence of breakthrough, defined as the destruction of cohesiveness of the defence, was an important event in the eventual success of an offence. Following breakthrough, 86% of operations were successful, whereas if no breakthrough was achieved only 15% were eventually successful. Once breakthrough has been achieved, it becomes possible for the attack force to conduct a type of operation more in the nature of exploitation than combat. Moreover, variations in the time to breakthrough also led to differences in the nature of campaigns; the timings of "immediate" (less than half a day), "quick" (less than 2 days), and "prolonged" (over 2 days) were used for study purposes.

In studying the course of operations, one of the aspects examined was the nature of the movements of attack forces. The study of the rate of terrain capture with operational success showed patterns that could be related to breakthrough time. Several different measures were examined, the first one derived having the dimensions of advance rate (distance/ time). Whilst the measures change through time, the following results relate to measurement (such as the length of attack frontage) at the time of breakthrough.

Area Taken and Mean Advance at Breakthrough

This measure may be taken as a possible index of how badly the defence has fared and of its chances of recovery. However, as the simple area measure does not allow for variations in size of operation, the function (area at breakthrough)/(attack frontage) has been used here. This has a dimension of length and would represent "mean advance" on the original attack frontage.

The distributions of the measures in Tables 3.1, 3.2, and 3.3 are log-normal within a given breakthrough category, as we will show explicitly later. In each case, the table shows the geometric mean of the corresponding log-normal distribution. The discussion of stock price returns and Brownian motion in Chapter 1 gives us some insight into why log-normal distributions might arise in such cases.

	Operations Leading To:	
	Campaign Success	Campaign Failure
Prolonged Breakthrough	29	8
Quick Breakthrough	5	-
Immediate Breakthrough	11	2.2

Table 3.1: Geometric Mean Area/Attack Front at Breakthrough (Miles)

	Operations Leading To:	
	Campaign Success	Campaign Failure
Prolonged Breakthrough	4.5	1.5
Quick Breakthrough	4.5	1.5
Immediate Breakthrough	11	2.2

Table 3.2: Geometric Mean Area Per Day/Attack Front at Breakthrough (Miles/Day)

	Operations Leading To:	
	Campaign Success	Campaign Failure
Prolonged Breakthrough	4.5	2.1
Quick Breakthrough	8.8	-
Immediate Breakthrough	18	5.5

Table 3.3: Geometric Mean √Area Per Day at Breakthrough (Miles/Day)

A first analysis of the results in Table 3.1 shows that these "mean advances" at breakthrough are greater for those achieving subsequent success than those failing by factors of 5 for immediate breakthrough and approximately 3.6 for prolonged breakthrough. This larger factor for immediate breakthrough is indicative of the extra "brittleness" that pertains to these very quick breakthrough cases. They also show that these "mean advances" at breakthrough are less for immediate than for prolonged breakthrough by mean factors of 0.38 for those achieving subsequent success and by 0.28 for those failing.

Moving to the more complicated measure of irruption in Table 3.2, we move from mean advance to mean advance per day, giving a first order measure of the mean rate of advance. For this case, there is somewhat less variability in results. There is, in fact, no significant difference between quick and prolonged breakthrough effects in this case. These two groups can therefore be pooled. Combined quick and prolonged breakthrough results in terms of subsequent success and failure are thus shown in Table 3.2.

We again find differences on this measure (mean rate of advance) between breakthroughs leading to eventual success and those leading to failure. These differences are by factors of 5 for immediate breakthrough, and 3 for prolonged/quick breakthrough. The mean rate of advance for immediate breakthrough is greater than for prolonged/quick breakthrough.

The caricature of movement to breakthrough implied is:

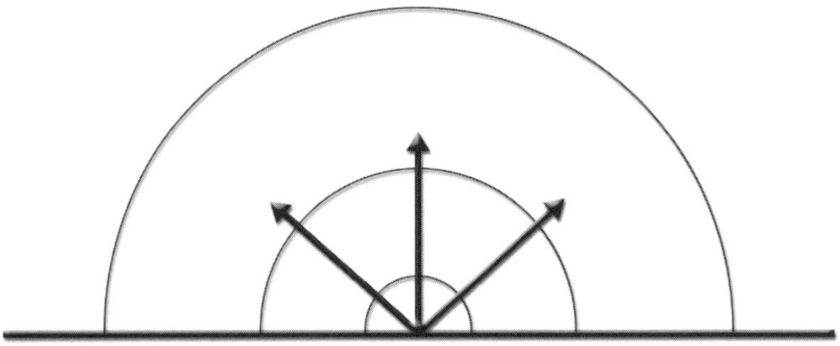

Rather than:

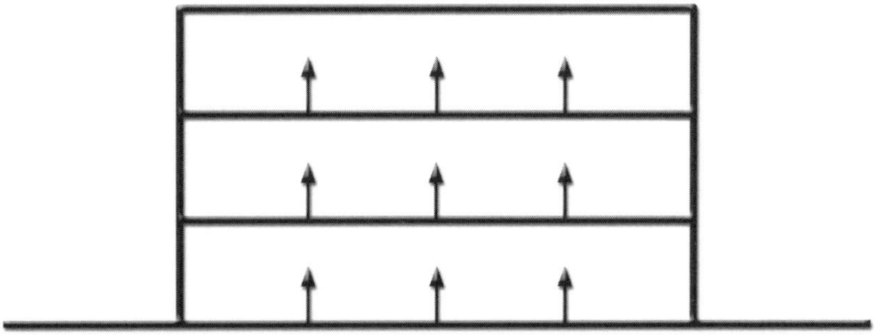

However, the movement *post-breakthrough* is better represented by the linear model than the radial one for success after breakthrough and without breakthrough. The exception to this is success after immediate breakthrough, which discriminates success most significantly on the basis of the radial model.

An alternative measure of the same dimensions is thus $\sqrt{\ }$ Area/Time, again a representation of advance rates but representing propagation from a point rather than a linear advance.

The equivalent mean values for this measure are shown in Table 3.3. The differences between groups show similar patterns to the previous measures, but are greater, and significant at the 1% level (*t* test) between immediate and prolonged breakthrough.

It can thus be observed that, despite the great variations in size of the operations studied, there are patterns to be deduced and these can offer lessons on the nature of this little-studied aspect of manoeuvre warfare.

This process of irruption has been identified as one of the key emergent effects of manoeuvre warfare [8]. We consider now whether such a process has scaling properties of the type discussed in our general consideration of complexity. The historical data indicates (as we have discussed) that for a given type of breakthrough (immediate, quick, or prolonged–I, Q, or P), and subsequent effect on the campaign (Subsequent Success [SS] or Subsequent Failure[SF]), the mean advance at breakthrough turns out to be a log-normal distribution. Of even more interest to us is the fact that if these distributions are plotted for each of the breakthrough/campaign effect categories, then they have a certain *scaling* character, which we now define.

Consider Figure 3.10 (page 71). Here we have plotted the log-normal distribution of mean advance at breakthrough for each of a number of breakthrough categories. The *x*-axis is the log of the mean advance at breakthrough (in miles), and the *y*-axis is such that a cumulative normal distribution of the *x* variable will give a straight line. We can see that the various categories of breakthrough produce cumulative curves that are parallel to each other, with one exception. However, this case has significantly less data than the other cases and we take this to be the cause of this deviation. The fact that these lines lie parallel to each other means the following: given two such curves, corresponding to x variables $x(1)$ and $x(2)$, there is a scaling variable λ (which depends on the two categories of breakthrough being considered) such that

$$\text{Log } x(1) = \lambda \text{ Log } x(2)$$

$$\text{i.e. } x(1) = x(2)^{\lambda}$$

and the distribution of mean advance at breakthrough coincides for the variables $x(1)$ and $x(2)^{\lambda}$. In this sense, we can say that $x(2)$ can be scaled by a power transformation so that its distribution collapses onto that of $x(1)$.

An alternative radial measure of irruption, as we have discussed, is √(area at breakthrough)/(days to breakthrough), with dimensions of miles per day. If this is plotted on the same basis as the previous figure, we again have evidence for a form of "scaling collapse" of the type discussed above (Figure 3.11). Moreover, the stability of the two sets of data indicate that there are (at least) two categories of emergent behaviour for irruption and subsequent campaign outcome–linear advance and radial propagation from a point.

Each data point in Figures 3.10 and 3.11 is a campaign outcome, classified in terms of immediate (I), quick (Q) or prolonged (P) irruption, and Subsequent Success (SS) or Subsequent Failure (SF). The key to the data points is given below:

■ I(SF) ● I(SS) □ Q(SS)

▲ P(SF) × P(SS)

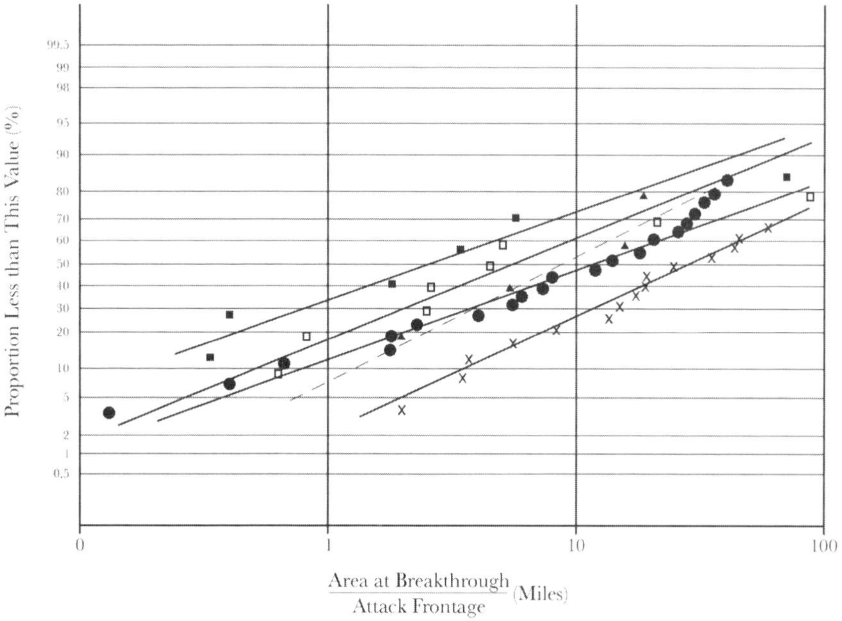

Figure 3.10: The Statistics of Linear Irruption

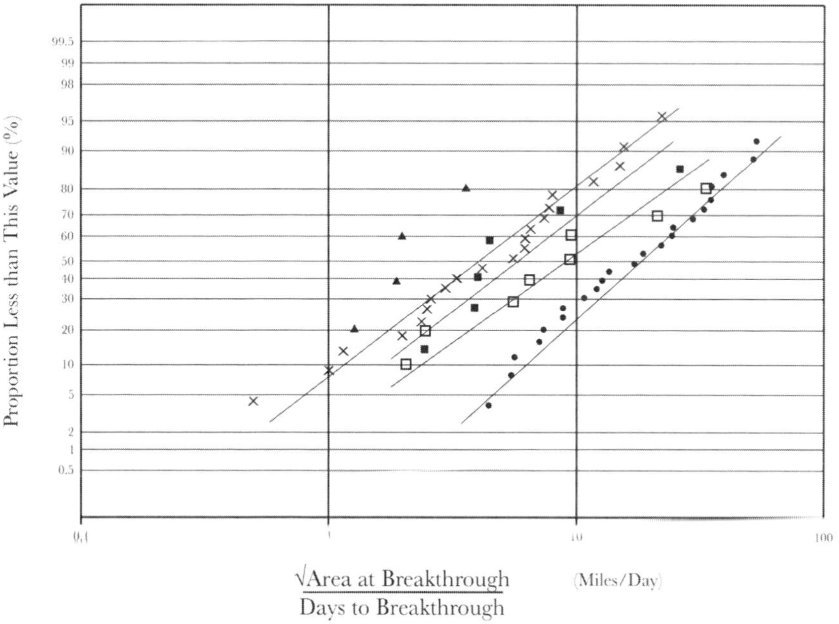

Figure 3.11: The Statistics of Radial Irruption

The Fractal Front of Combat

In [9], Lauren discusses the fractal nature of a combat front between two opponents. The idea is that an essentially straight line frontage between two tactical-level opponents will buckle into a fractal shape, whose fractal dimension can be calculated as a function of the force ratio of the forces involved (the number of attackers to the number of defenders) as derived from Historical Analysis of infantry battles carried out by the UK Dstl.

Lauren [9] uses the Historical Analysis result as a basis for his analysis:

F = (number of attack infantry/number of defence infantry)$^{0.685}$

where F is a multiplier for the base number of casualties of the attacking force per defence weapon. As a consequence of this, Lauren was able to show that the combat front will buckle over time and in the limit will have a fractal dimension $D = 1.685$.

From Chapter 1, if we assume that this process is akin to invasion percolation of one fluid by another in a porous medium, the fractal dimension of the boundary of the resulting interface should lie in the range 1.33-1.89, which is what we find from historical data.

It is possible in this case to derive the underlying dynamics producing this statistical effect. It turns out that this fractal factor is due fundamentally to detection of targets [10], and comes from a model of the engagement process that leads to the following relationship:

$$\frac{1}{R} = \frac{k_1}{T} + k_2$$

where k_1 and k_2 are constants, R is the defender rate of fire, and T is the number of targets in view [11].

It reflects the asymmetry of the infantry battle in the following sense [12]. The attack force aim is to close on the defence position, and fire is used in a general suppressive mode–actual casualties caused to the defence are only a small part of the process at this point. However, from the defence perspective, the aim is to deter the attack, and casualties to the attack force are very important. Such casualties to the attack force are a direct reflection of the intervisibility of targets to the defence force as discussed above.

As with most applications of fractal processes, the process breaks down at some point due to the granularity of the resolution. In this case, the process remains valid up to about 30m closing distance between the attack and defence. At that point, a different mechanism comes into play, leading to local defence surrender and attack overrun of defence positions [12].

More generally, the figure of 0.685 relates to open terrain. In urban areas it is about 0.5 [13]. The closing to overrun appears to occur differently in urban and wooded terrain as compared with open terrain [12]. For example, in open conditions, the closing part of the battle occurs across the front. By contrast, in urban conditions, the attack force is split into small subunits who individually close on defence locations leading to local surrender and overrun.

Power Law Relationships in Combat Data

The Hartley Model

In [14], Hartley has analysed eight separate databases of historical combat data. Four were developed by Helmbold; the fifth dataset, "Inchon," was developed by Busse. The last three datasets were developed by Dupuy. These eight datasets span several centuries in time, include both air and land conflicts, and span the range from small to large interactions. Details of the databases are given in [14]. On the basis of this extremely extensive set of data, Hartley was able to develop a stable analysis of the relationship between casualties in conflict and the initial force ratio based on earlier ideas of Helmbold. He defines the following two dimensionless variables:

$$\text{HELMRAT} = \frac{x_0^2 - x^2}{y_0^2 - y^2}$$

$$\text{FORRAT} = \frac{x_0}{y_0}$$

where in each case, x_0 is the starting value of force size, and x is the final value (similarly for y). Hartley has established a power law relationship between these two variables, HELMRAT and FORRAT, on the basis of the comprehensive data sets described above. He has shown that (in logarithmic terms):

$$\text{Ln (HELMRAT)} = \alpha \text{ Ln (FORRAT)} + \beta$$

where the expected value of α is approximately 1.35 and the value of β is approximately normally distributed about the value -0.22 with standard deviation of 0.7. Hartley shows that the value of α has the characteristics of a universal constant, being stable over four centuries of time [14, Figure 17], and stable when considering conflicts of different sizes, ranging from force sizes of less that 5,000 to more than 100,000 [14, Tables 4, 5, and 6].

If it is assumed that the mechanism that produces this remarkably stable relationship between casualty effects and force ratio is of Lanchester type, then Hartley shows that it must be of linear-logarithmic form. However, the relationship is based on the empirical data alone, and other explanations are possible. For example, in [15] an analysis based on self-organisation (in particular the forest fire model of Chapter 2) is put forward as the basis for the equally remarkable scaling of conflict size. It is thus persuasive that such complexity-based effects, rather than a Lanchester process, lie at the base of the scaling relationship established by Hartley. This analysis by Turcotte and Roberts is next discussed.

In their paper [15], they begin by comparing the predictions of the theoretical self-organising forest fire model with the statistics of the relative sizes of real forest fires. Four data sets are considered: 4,284 forest fires in the USA Fish and Wildlife Service Lands during the period 1986-1995; 120 of the largest fire areas in the western USA from tree ring data, spanning the period 1155-1960 (800 years); 164 fires in the Alaskan boreal forests during 1990-1991, and 298 fires in the Australian Capital Territory during 1926-1991. The results are in good agreement with a power law statistical distribution of size of fire versus frequency, with a power law exponent of between 1.3 and 1.5. The remarkable thing is the stability of the trend across such a long period of time, during which technology has changed, as have ways of fighting such fires. The authors then show that a similar power law relationship (also with an exponent in the same range) holds for the intensity of conflict versus its frequency. This work extends the research of Richardson [16], who also showed a power law relationship between the intensity of war and its frequency. Turcotte and Roberts base their results on two data sets: that of Levy, which tabulates the intensities of 119 wars from 1495 to 1973; and that of Small and Singer, who considered 118 wars over the period 1816-1980.

The similarity of the power law exponent for both forest fire statistics and war intensity leads Turcotte and Roberts to hypothesise that war deaths are caused by a self-organising mechanism akin to that of the forest fire model. This is at least the beginning of an explanation of why casualties in war should give rise to such a simple power law relationship, stable over centuries of time.

REFERENCES

1 LTG TUKER F (1948). *The Pattern of War.* Cassell. UK.

2 MOFFAT J (2002). *Command and Control in the Information Age: Representing its Impact.* The Stationery Office. London, UK.

3 KUHN G W S (1989). "Ground Force Casualty Patterns: The Empirical Evidence." Report FP703TR1.

4 WAYLAND R, BROMLEY D, PICKETT D, and PASSAMANTE A (1993). *Physics Review Letters.* 70. p. 580.

5 LAERI F (1990). *Computational Physics.* 4. p. 627.

6 LAUREN M and STEPHEN R T. "Fractals and Combat Modelling: Using MANA to Explore the Role of Entropy in Complexity Science." Paper prepared for *Fractals.* Defence Technology Agency. Auckland, New Zealand.

7 ROWLAND D, KEYS M C, and STEPHENS A B (1994). "Breakthrough and Manoeuvre Operations (Historical Analysis of the Conditions for Success) Annex I, Irruption." Unpublished DOAC Report, Annex I.

8 ROWLAND D, SPEIGHT L R, and KEYS M C (1996). "Manoeuvre Warfare: Some Conditions Associated with Success at the Operational Level." *Military Operations Research.* 2 No 3. pp. 5-16.

9 LAUREN M K (2000). "Modelling Combat using Fractals and the Statistics of Scaling Systems." *Military Operations Research.* 5 No 3. pp. 47-58.

10 ROWLAND D (1983). "The Effectiveness of Infantry Small Arms Fire in Defence – A Comparison of Trials and Combat Data." Unpublished DOAC Memorandum.

11 THODY J H and DOVE H J (1981). "An Analysis of Small Arms Fire by Infantry in Defensive Positions." DOAE Unpublished Note.

12 ROWLAND D personal communication.

13 ROWLAND D (1991). "The Effect of Combat Degradation on the Urban Battle." *Journal of the OR Society.* 42 No 7. pp. 543-553.

14 HARTLEY D S (1991). *Confirming the Lanchesterian Linear-Logarithmic Model of Attrition.* Martin-Marietta Center for Modelling, Simulation and Gaming. Report K/DSRD-263/R1.

15 ROBERTS D C and TURCOTTE D L (1998). "Fractality and Self-Organised Criticality of Wars." *Fractals.* 6 No 4. pp. 351-357.

16 RICHARDSON L F (1960). *The Statistics of Deadly Quarrels.* Boxwood Press. Pittsburg, USA.

CHAPTER 4

MATHEMATICAL MODELLING OF COMPLEXITY, KNOWLEDGE, AND CONFLICT

INTRODUCTION

Understanding the behaviour of agent-based simulation models of conflict is now becoming more important, especially as (with improved representation of Command and Control [1]) the agents gain intelligence and try to outsmart each other, producing potentially very complex behaviour. In the modelling of natural systems (such as fluid dynamics, or heat flow), the principal variables in these models can often be separated out from the rest of the model to produce a mathematical metamodel that is aimed at relating the outputs of the model to these driving inputs in a more transparent and explicit way. If this can be achieved, it improves our understanding of the system and its likely emergent behaviour, as well as complementing the use of detailed simulation. Such an approach is consistent with the idea of 'Operational Synthesis' as espoused by Dr. Alfred Brandstein, then Chief Scientist, U.S. Marine Corps [2].

As an example, in developing a metamodel we consider the relationship between a key outcome of the model, a, and a set of input variables as follows:

$$a = f(a_1, \ldots a_k, b_1, b_2)$$

(This is easily generalised to an arbitrary number of *b*s.) The arguments $a_1, \ldots a_k$ have independent dimensions. That is, the dimension of any a cannot be expressed as a combination of the dimensions of the other *a*s. In contrast, the dimension of each b variable can be expressed as such a combination. The arguments can be transformed using a gauge transformation so that:

$$a_1' = A_1 a_1, \ldots, a_k' = A_k a_k \, .$$

These correspond to a change in the "gauge" (e.g., from centimetres to metres or kilometres) in the measurement of a variable. Physically, if a gauge change makes no difference to observed behaviour for all observers using different gauges, the variable is said to be *self-similar.*

The metamodel function can then be shown [1] to have the property:

$$a = f(a_1,...,a_k,b_1,b_2) = a_1^p...a_k^r\Phi\left(\frac{b_1}{a_1^{p_1}...a_k^{r_1}},\frac{b_2}{a_1^{p_2}...a_k^{r_2}}\right)$$
$$= a_1^p...a_k^r\Phi(\Pi_1,\Pi_2) \qquad (1)$$
$$\text{where } \Pi_1 = \frac{b_1}{a_1^{p_1}...a_k^{r_1}} \text{ and } \Pi_2 = \frac{b_2}{a_1^{p_2}...a_k^{r_2}}$$

These "Π" variables are sometimes called *similarity* variables. This is because two natural systems, with different values of *a*s and *b*s but the same value of Π, will tend to have similar emergent behaviour. An example is the flow of air past an aircraft in the atmosphere or past a model of the aircraft in a wind tunnel. The Π variable in this case is the Reynolds number. If this is the same in both cases, then the model in the wind tunnel will give results relevant to the full-scale aircraft in the atmosphere.

Self-similar solutions correspond to problems where the values of the variable (b_2 for example) tend to zero or infinity. Three possibilities are available (see Reference [1] for further discussion):

TYPE 1 METAMODEL. Φ tends to a non-zero finite limit as one of its arguments tends to zero or infinity. This means that in most practical cases, this argument can be eliminated from the relationship, giving a simplified form in equation (1).

TYPE 2 METAMODEL. Φ has power-law asymptotics of the form:

$$\Phi(\Pi_1,\Pi_2) = \Pi_2^{\alpha_1}\tilde{\Phi}\left(\frac{\Pi_1}{\Pi_2^{\alpha_2}}\right) \qquad (2)$$

as the argument Π_2 tends to zero or infinity. In theoretical physics, these systems are examined from a gauge theory point of view using a "renormalisation group" approach in which the parameter Π_2 is considered (through the repeated application of a renormalisation group) at larger and larger (or smaller and smaller) gauges, giving an asymptotic expression of the form required for a type 1 or 2 metamodel. Later, we will discuss the renormalisation group in more depth and relate it to concepts of control of the battlespace.

TYPE 3 METAMODEL. Neither 1 nor 2 holds and self-similarity is not observed; Φ has no finite limit different from zero and no power-law asymptotics.

This approach directs us (for evidence of metamodels of types 1 and 2) to search for evidence of power-law relationships of the form $y = x^{\alpha}$, which, if plotted on a log-log scale, give a straight line whose slope is the power-law exponent. Such expressions arise naturally in certain types of complex systems, particularly where fractal structures are involved, and are referred to as *scaling relationships* since they have no preferred gauge. *Evidence of such scaling relations is thus evidence in support of the assumption of relative gauge.* Chapter 3 shows that there is clear evidence for such an assumption in historical conflict data, and that metamodels of types 1 and 2 should be expected for agent-based models of conflict. In addition, we should expect to see evidence for normalised "scaling collapse" as exemplified by the function $\Phi(\Pi_1, \Pi_2)$ in equation (2), and we should expect to see the effect of renormalisation groups.

If we consider an agent-based "distillation" such as the ISAAC model developed under Project Albert by the U.S. Marine Corps Combat Development Centre, we can consider the emergent behaviour of such a model in terms of both the spatial clustering of the agents, and the attrition that they inflict on the opponent. For such a distillation, a metamodel of type 2 applies [1] that allows us to relate the attrition rate for one side to the clustering dynamics of the opposing side, as measured by the mean fractal dimension of these clustering agents. As a simple example, (given in [1]), assume that the command process, say for Red, is represented by the following effects:

1. The number of discrete clusters of Red agents at time t, $N(t)$, is specified ahead of the simulation.
2. $N(t)$ is a decreasing function of t.

These assumptions are meant to suggest that the number of Red clusters decreases in time, reflecting the desire of Red to concentrate force. With these assumptions let us further assume that the smallest cluster of Red agents, $X(t)$, at time t, is taken and added to another randomly chosen cluster of Red agents. This process thus represents both the concentration of Red force and the reconstitution of force elements.

Let us now define $\varphi(x,t)$ = (expected number of clusters of Red agents $\geq$ size x at time t)/(initial total number of clusters of Red agents) and $N(t)$=(the total number of remaining clusters of Red agents at time t)/(initial total number of clusters). Given the assumptions and definitions above, it can then be shown that $\varphi(x,t)$, the cumulative distribution of cluster sizes at time t, approaches a self-similar distribution as time progresses (i.e., a scaling collapse takes place). Thus the cluster size distribution evolves over time by a scaling relation. $\varphi(x,t)$ can then be represented in the self-similar form:

$$\varphi(x,t) = \frac{g(x/X(t))}{X(t)}$$

where g is some positive continuous function and where $g(1)=N(t)X(t)$. This self-similar form means that we can define the distribution of relative cluster size in a way that is time invariant (although the actual cluster sizes will change).

Now assume that the evolution of the distribution of $\varphi(x,t)$ is smooth (a small change in time t leads to a small change in $\varphi(x,t)$; this is equivalent to saying that the renormalisation group is smooth [1]). If $\overline{\varphi}(x,(1-\delta)t)$ is the expected cluster size x at time $(1-\delta)t$ and $\overline{\varphi}(x,t)$ is the same expectation at time t, then this assumption means we can find a constant b to first order such that:

$$\overline{\varphi}(x,t)=(1+b\delta)\overline{\varphi}(x,(1-\delta)t).$$

It then follows that:

$$\log\overline{\varphi}(x,t)=b\log t+c$$

for some constant c and the normalised expected cluster size at time t, $\overline{\varphi}(x,t)$, varies as a power-law with increasing time t and scaling constant b.

If ΔB is the change in the number of Blue agents, Lauren [3] has shown that:

$$\frac{\Delta B}{\Delta t}$$

is proportional to the product of (Red unit effectiveness)x(the probability of meeting a Red cluster)x(the expected number of Red units per cluster). It is assumed that unit effectiveness is constant. Keeping the cluster size constant for the moment, this indicates [3] that the rate of change of Blue agents is given by an expression of the form:

$$\left\langle \frac{\Delta B}{\Delta t} \right\rangle = k^{q(D)} \Delta t^{r(D)}$$

where D is the average fractal dimension of Red (and therefore an indication of how Red clusters/collaborates locally) and both r and q are exponents. This equation is a form of Lanchester law where the rate constant is dependent upon the clustering of Red agents. If Red cluster size varies according to:

$$\varphi(x,t) = \frac{g\left(x / X(t)\right)}{X(t)}$$

where x is the cluster size and $X(t)$ the smallest cluster at time t, then we can write:

$$\left\langle \frac{\Delta B}{\Delta t} \right\rangle = k^{q(D)} \Delta t^{r(D)} N(t) \bar{g}(y(t))$$

where $N(t)$, inversely related to $X(t)$, is the normalised number of clusters of Red at time t and $\bar{g}(y(t))$ is the mean of the distribution of cluster size, which evolves as a power law (as we have shown in certain cases).

More General Distribution of Cluster Sizes

For self-organising groups of agents that are approaching a critical point in the form of a Bak-Sneppen evolution model, but are not necessarily attempting to concentrate force or to reconstitute force in the sense described above, our previous analysis in Chapter 1 indicated that the distribution of cluster sizes at some intermediate point s is given by:

$$P(S, f_{i,s}) = S^{-\tau} g(S(f_c - f_{i,s})^{1/\sigma})$$

where $f_{i,s}$ is the minimal site probability at timestep s. As noted before, this converges to a power-law distribution as s tends to infinity.

In [1], we also relate the cluster fractal dimension predicted by the Bak-Sneppen evolution model of local species coevolution, to that which emerges from initial experiments with the ISAAC cellular automaton model. (For a description of ISAAC see the Web site at reference [4].)

The Renormalisation Group

The gauge invariant approach to metamodelling outlined above also leads us to consider the role of the renormalisation group, which explicitly appears in terms of its effect on the distribution function $\Phi(\Pi_1, \Pi_2)$ in our characterisation of metamodels. Let us explore this a little more here. Suppose that $u(x,t)$ is a function of the two variables x and t. Using the "static scaling" assumption of renormalisation [5], we assume that we have a group of renormalisations of the form $R_{b,\phi}$, so that:

$$R_{b,\phi}u(x,t) = \frac{1}{Z(b)}u(b^{\phi}x, bt)$$

From the group property, $R_{a,\phi}R_{b,\phi} = R_{ab,\phi}$

It follows that $Z(a)Z(b) = Z(ab)$

Thus $Z(b) = b^{\alpha}$ for some exponent α

If $u^*(x,t)$ is a fixed point of the renormalisation, then;

$$u^*(x,t) = b^{-\alpha}u^*(b^{\phi}x, bt) \quad \forall b$$

Choose $b = \frac{1}{t}$, then;

$$u^*(x,t) = t^{\alpha}\tilde{u}^*(\frac{x}{t^{\phi}})$$

The repeated application of the renormalisation process will thus (in the limit) produce a functional relation of the form u^*. This explains why the type 2 metamodel has the form assumed, and also why the (renormalisation invariant) cluster size distribution of Carr and Pego discussed earlier has the form derived by them. Note the simplification that has been achieved in going from an unknown function of two variables to a normalised unknown function of only one variable.

The ability of a force to control an area of operations can be related to the fractal dimension of the force through the use of such a renormalisation process as we now show.

Control of the Battlespace

Looking now at the phenomenon of control of the battlespace, we can consider the problem using the renormalisation approach as in a type 2 metamodel. Consider, as shown in Figure 4.1, the Area of Operations (AO) of a military commander. For simplicity we assume this is a square of side L.

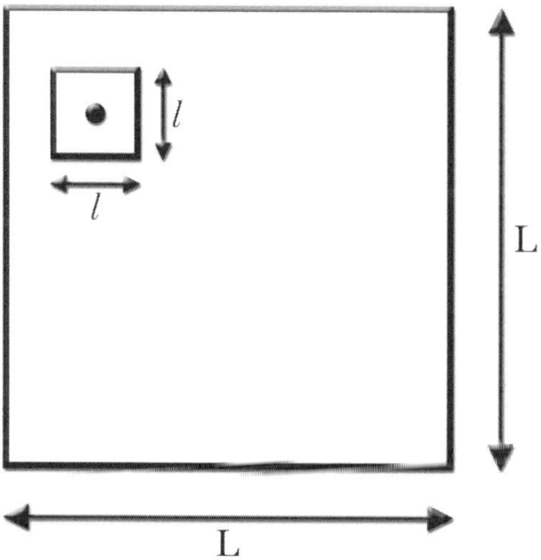

Figure 4.1: Area of Operations

We assume that the commander aims to establish control in this area. Firstly we have to define what this means. Each unit, shown by a dot in Figure 4.1, has an area surrounding it that it can control. The size of this area is defined by the nature of the force and its associated sensors [6], giving rise to a "bubble" of control around the unit. In two dimensions, let this area correspond to a square of side l. We assume that l is significantly smaller than the dimensions of the AO. (Note: we have restricted the battlespace here to two dimensions to simplify the discussion. However, the same approach should work in three dimensions, corresponding to the complete battlespace.)

Now let D be the fractal dimension of the force under the commander's control within the AO. Suppose we partition the AO into square cells of width l. Let N be the total number of such cells, so that $Nl^2 = A$. Let $N(l)$ equal the number of cells in the AO that are occupied by one of the units making up the force. By definition of the fractal dimension, we have that $N(l) = l^{-D}$ (normalising the constant of proportionality to 1). If p is the probability that a cell chosen at random in the AO is under control, then:

$$p = \frac{N(l)}{N} = \frac{l^{-D}}{N} = \frac{l^{2-D}}{A}.$$

Note that D always lies between 0 and 2, so that p is well defined.

In discussion with senior UK commanders who have had recent operational experience at a high level, the concept of *control* of an area as corresponding to the prevention of flow through an area (flow in terms of an opposing force, or perhaps some third party) has been endorsed as a good analogy. We thus define the commander as having "weak control" of his area of operations if he can to some extent control movement through the AO. This is similar to the problem of determining whether fluid can seep through a block of semi-porous rock, as discussed in [7] where a renormalisation group approach was used. We thus define weak control as corresponding to a span of controlled areas that stretch either from side to side or top to bottom of the AO. Following on from this, we define the commander to have "strong control" of the AO if there is a span of controlled areas stretching *both* from side to side *and* top to bottom of the AO, resulting in a strong constraint on the flow of people through the area.

The question at issue is then: how do these concepts of control relate to the ability of the force to collaborate locally (the fractal dimension)?

Consider first a cell of four elements where each cell is a square of side l, which a single unit can control. We now consider the probability p_1 of weak or strong control of this square cell of side $2l$ in terms of the probability:

$$p = \frac{l^{2-D}}{A}$$

of a unit controlling each of the squares of side l. We consider each of the five different classes of configuration for this cell, as shown in Figure 4.2.

In Figure 4.2, we show the five classes a to e of configuration, and mark beside each case whether this gives weak or strong control, by considering the span of controlled areas.

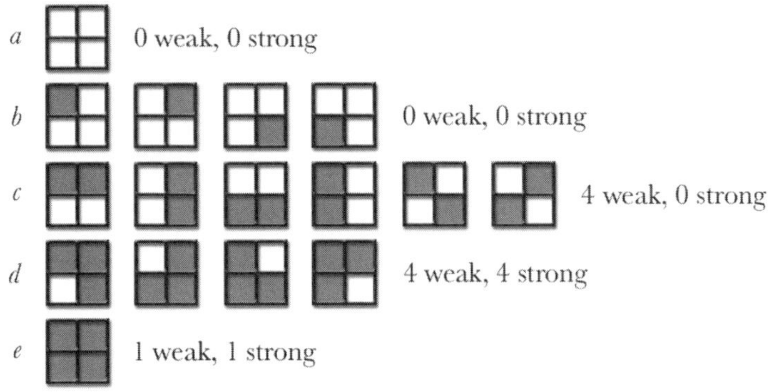

Figure 4.2: Five Configuration Classes

The probability of each configuration can be derived in terms of p. For example, the probability of any of the cases in configuration d is $p^3(1\text{-}p)$. By adding up the configurations corresponding to weak control, and taking into account the probability of each such configuration, we have the relation:

$$p_1(\text{weak}) = 4p^2(1-p)^2 + 4p^3(1-p) + p^4.$$

We can do the same thing for strong control, leading to the relation:

$$p_1(\text{strong}) = 4p^3(1-p) + p^4.$$

Using the renormalisation group approach [5], we iterate at increasing levels of cell size, leading to the relations:

- Weak control: $p_{n+1} = p_n^2(4 - 4p_n + p_n^2)$
- Strong control: $p_{n+1} = p_n^3(4 - 3p_n)$

These give rise to the recursive schemes shown in Figures 4.3 and 4.4. The relations for weak and strong control above correspond to the relationships respectively:

$$f(x) = x^2(4 - 4x + x^2)$$
$$g(x) = x^3(4 - 3x)$$

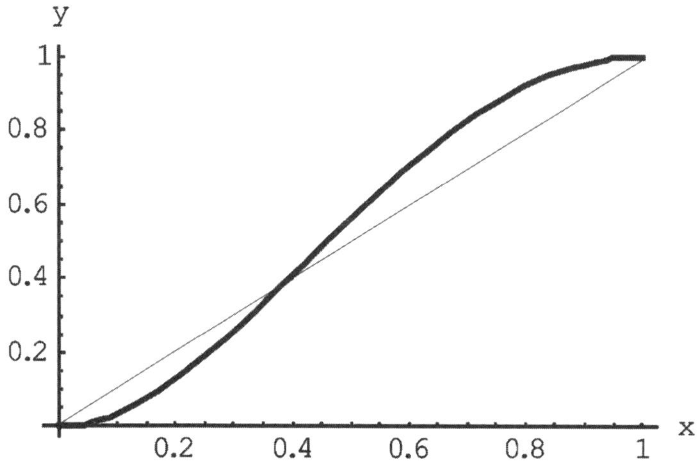

Figure 4.3: Plot of $y = f(x)$ and $y = x$ - weak control

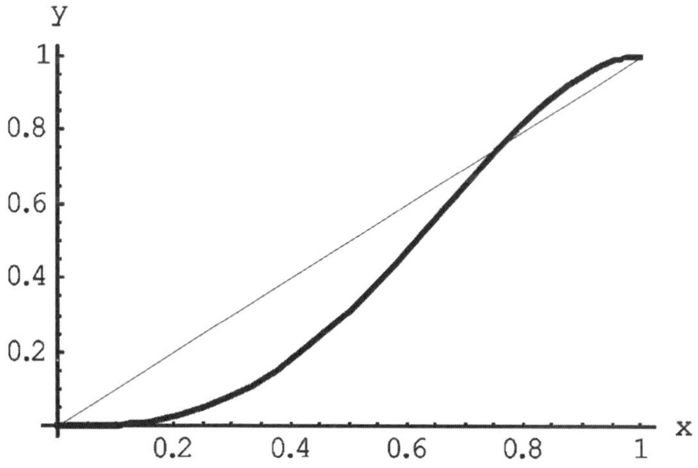

Figure 4.4: Plot of $y = g(x)$ and $y = x$ - strong control

The fixed points in the recursive relation of weak control correspond to the values x shown in Figure 4.3 such that $y = f(x)$ intersects $y = x$. Similarly for strong control, the fixed points correspond to the values x such that $y = g(x)$ intersects $y = x$. For both weak and strong control, there are stable fixed points at $x = 0$ and $x = 1$. However, there is also an unstable fixed point between these that is different for strong and weak control. This was calculated to be 0.382 for weak control, and 0.768 for strong control.

Control and Fractal Dimension

In either case, starting with a given fractal dimension for the force, and the dimensions of the AO, we can calculate a corresponding starting probability:

$$p_0 = \frac{l^{2-D}}{A}.$$

For side length L of the AO there will be a corresponding value of iteration order n such that $2^n l \cong L$. By using the recursive scheme above, we can calculate for this value of n the corresponding probability of weak or strong control of the AO. Consideration of Figures 4.3 and 4.4 indicates that there is a critical value of the probability:

$$p_0 = \frac{l^{2-D}}{A}$$

such that values above p_0 polarise towards very good control, whereas values below p_0 polarise towards very poor control. In fact, the point p_0 corresponds to a phase change in the behaviour of such a system.

Examination of Figures 4.3 and 4.4 shows that it is easier to iterate towards good weak control than towards good strong control, as we would expect (since weak control is easier to achieve than strong control). Figure 4.5 shows how this iteration works for a starting probability of 0.65 and the requirement of weak control.

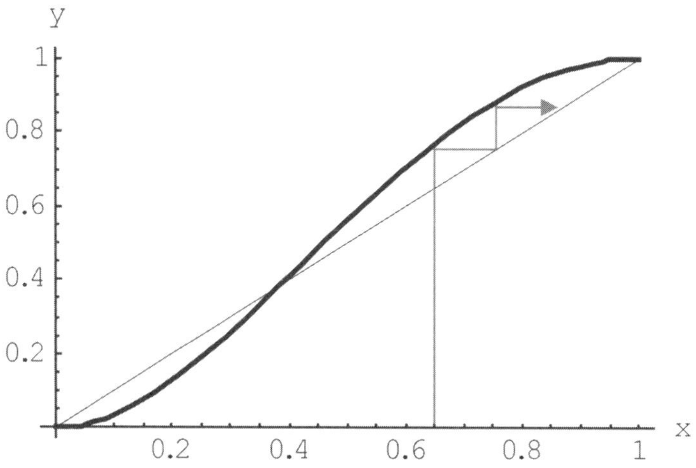

Figure 4.5: Recursive Calculation of the Probability of Weak Control

Lockout

From a game theoretic perspective, we can see that each side is trying to drive its own value of control up, and the other side's down. The analysis above indicates that there should be rapid *lockout*, i.e. one side should rapidly gain control and lock the other side out.

Percolation Theory and the Renormalisation Group

In relating these ideas to the behaviour of natural systems, the various configurations of 2x2 controlled areas are identical to the porous and nonporous regions in semipermeable rock structures [7]. The study of such processes is referred to as *Percolation Theory*. A good step-by-step introduction to the theory and some working examples of how such processes work in two dimensions are provided at the Web site reference [8]. If p is the probability of an individual rock (or crystal) domain being porous, then p is the driving parameter of the process. The behaviour can be tuned in the sense that we discussed in Chapter 2. If p is below the critical value p_0, then the clusters are not large enough to form a path of percolation from one side of the structure to the other (popping noise). When the probability p is above the critical value p_0, then suddenly all clusters tend to spread from one side to the other, allowing percolation throughout the structure (these are thus called *percolation clusters*) and corresponding to a phase change in the dynamic of the system (snapping noise). Near to the critical point, it can be shown [8] that the distribution in size of percolating clusters is a power-law, corresponding to a fractal distribution of cluster size (crackling noise). In fact, for p slightly greater than p_0, the fraction F of individual domains that form part of a percolating cluster takes the form $F = F_0(p - p_0)^\beta$. From this, we can see that as p approaches the critical value p_0 from above, the fraction of domains forming part of the percolating cluster tends to zero (for a very large initial configuration). Thus such clusters can become very tenuous close to the critical point.

These configurations come originally from attempts to model lattices of magnetic spins in more than one dimension [5]. Such 2x2 configurations are then referred to as *block spins* since they are composed of four individual spins, each of which may be either up or down (the block spin is defined as the sum of the signs of the individual spins, so it still has the value +1 or −1). By developing the idea of the *renormalisation group*, iterating to larger and larger domains, Wilson (building on work by Kadanoff), was able to show that such arrays of magnetic spins do indeed exhibit phase change effects (as we have shown above for the phase change between being out of control and being in control of a region), and that the parameters involved can be calculated explicitly. For this work, Wilson was awarded the Nobel Prize for Physics.

Implications for Self-Organising Information Networks

If a Self-Organising Information Network is thought of as a grid of connections, and we have a probability p of creating a link between one element of the network and another, then percolation across the network corresponds to being able to send a signal from one end of the network to the other. From Percolation Theory, we can thus see that we would expect a phase change in the dynamic of such a network. If p is small, then only local connections can be made. However, at some critical value of p, there will be a phase change such that connections across the full network can suddenly be made. Near to this critical value of p, the clusters formed by those connected on the network will form a fractal set, and the distribution of such cluster sizes is described by a power-law relationship between size and frequency of that size. Two questions that arise are:

1. What is the benefit (and cost) of being above the critical threshold so that the connections are robust?
2. How can we measure the benefit of using the knowledge obtained by such networked interconnection?

Recent work by Perry [9] has exploited the idea of information entropy to address the second question (with a reduction in entropy across the network corresponding to an increase in knowledge, and this then being equivalent to a reduction in delay in prosecuting an action). To understand where this idea originates, we first look at the influence of knowledge in wargames from an open systems perspective.

Wargames as Open Systems Sustained by Knowledge Flowing Across the Boundary

As noted recently by Roske [10], a wargame is an open system of the type introduced in Chapter 1. As he notes,

> *"In a classic command post exercise, we inject human decision-making into a structured environment to generate open system behaviours. Human decisionmaking represents energy crossing the structured system boundary…"*

In [11], Perry and Moffat developed the idea of using Information Entropy (from Shannon Information Theory) as a way of capturing the knowledge available to a military commander's decisionmaking during a wargame. This was initially applied to looking at the benefit to be gained from advanced airborne standoff radars (such as JSTARS in the U.S. or ASTOR in the UK).

A series of wargames was carried out in the UK to quantify the difference in combat effectiveness of a force without airborne standoff radar (ASTOR) in comparison with a force with ASTOR, or a force with other weapon systems whose life-cycle costs approximately equalled those of ASTOR. The common thread through all of the cases to be examined was the stream of decisions made by the friendly force commander. We describe here how we were able to capture the flow of information to the Blue commander during the wargames using the concept of Information Entropy, and then turning that into a measure of "knowledge." This approach allowed us to measure the quantity of knowledge flowing across the boundary of the system in order to influence the decisions made by the Blue commander and the onward evolution of the wargame.

Wargame Structure

There are three major problems with the use of wargames to support military studies: (1) too little output data, (2) the likelihood of atypical results, and (3) oversimplification. The first problem stems from the fact that wargames are generally slow, cumbersome, and resource intensive. Consequently, most analysts who use them to support studies plan only a small number of games, thus precluding significant statistical results. The second problem recognises the possibility that the sequence of decisions taken by the players in these games represents *statistical outliers*. Players may adopt extreme strategies that exist "outside" of what is considered to be a typical military response. The third problem reflects the fact that human players can only approximate the results of combat operations. In our studies, we addressed these problems in three ways: by arguing that our wargames are *quasi-memoryless processes* for tactical situation assessment; by introducing the *epitomising strategy principle* in wargames; and by embedding computer models to adjudicate engagements in the manual games. We discuss the first two of these concepts next.

The Memoryless Property of Wargames

The wargames used to support this study were two-sided, zero-sum games played over several discrete time periods or cycles (Bowen [12]). Consequently, the entire campaign can be viewed as a dynamical process in which the state variables are the force levels on both sides. In each of the wargames, both the friendly and enemy commanders formulated operations plans based on the stated campaign objectives. During the play of each game, both commanders assessed the overall situation at the strategic and tactical levels. Assessment at the strategic level consisted of examining the need to alter the campaign plan. At the tactical level, it supported force allocation decisions consistent with the implementation of the campaign plan.

A dynamic process is said to be *memoryless* or *Markovian* if at each cycle, the state of the system is influenced only by the state of the system in the previous cycle, and not by the specific history of the system (Stark and Woods [13]). The wargames conducted to support this study *approximately* satisfied these conditions as follows:

- TACTICAL DECISIONS: In all of the wargames played, the strategic situation was such that neither commander found it necessary to alter their original plan. Consequently, the commanders' decisions centred on the allocation of their forces. This forced them to focus exclusively on the tactical situation in the current game cycle and their assessments of the likely situation in subsequent cycles. Transitions in the state variables therefore depended upon the status of the forces at the end of the previous cycle, the decisions taken in the current cycle, and the combat attrition experienced in the current cycle.
- CYCLE INDEPENDENCE: Both commanders were assumed to act outside the opponent's decision cycle. That is, the opponent was assumed to have sufficient time within each cycle to redeploy his forces so that intelligence on unit identity, type, and location gained in the previous cycle was no longer valid in the current cycle; past history had no effect on the commanders' current

> decisions. In actual practice, we found this assumption only partially valid, as will be made more apparent below in the discussion of the UK FASTHEX gaming model used in the wargames for this study.

Under these conditions, the campaign can be viewed as a sequence of decisions taken in a discrete dynamical system. Each commander attempts to select a set of decisions (strategy) that will maximise the likelihood that he will achieve his mission and that is consistent with the overall campaign plan. Because the decisions are made under a connecting campaign plan, we assert that the process is only *quasi-memoryless*. This assumption is most relevant when the situation on the ground is in a state of rapid flux–the most interesting case.

The Epitomising Strategy Principle

In the play of the games, we strove to ensure that the commanders took actions that epitomised the side's historic conduct in battle. We attempted to avoid bold, daring, brilliant manoeuvres as well as bungling incompetence. For the purposes of analysis, more cautious conservative strategies that are consistent with accepted doctrine are preferred. There is a real danger that uncontrolled play would have resulted in producing only *outlier* results, when what is wanted are *typical* results. We achieved this by playing the wargames "open."[1] Red and Blue players were able to discuss strategies and decisions in the presence of a game controller and thus we ensured that the actions contemplated by either did not constitute extraordinary tactics.

---

[1]A wargame is "open to Blue" if the physical state of Red is fully known to the Blue player (Bowen [12]).

In keeping with this principle, great pains were taken to ensure that each side "knew" only those things that would normally be known through the available surveillance and intelligence assets. Players were forced to examine the information received through sensors and surveillance assets carefully to ensure that:

1. Sound military judgement was used in considering the decision options available.
2. Players used what they learned and interpreted from sensor reports, and not what they saw on the "game board."
3. The information received was consistent with the limitations and capabilities of the equipment being used, the employment of the surveillance assets, and other environmental conditions.

Wargaming with FASTHEX

In the play of the FASTHEX wargame, the continuity of the battle was modelled as a series of discrete timesteps referred to as "game cycles" (Figure 4.6). The length of a cycle can be set by the players, but is usually chosen to be 2 hours. Within each cycle, a linear sequence of actions is taken and the consequences of each is evaluated to simulate events within the cycle. Blue and Red alternately begin the sequence in order to smooth out the advantages or disadvantages of "going first."

The flow chart in Figure 4.6 can be thought of as a series of modules in a fully automated simulation of combat, less the decision module. The modules are rather simple representations requiring extensive player input. For example, the player selects the type of reconnaissance system to be used, states the current environmental conditions, and the hexagon[2] on the game board to be searched. With this information, the model applies the appropriate probability of detection or recognition and reports the results. Environmental conditions and terrain features are not known by the model and are inserted through rules constraining the play.

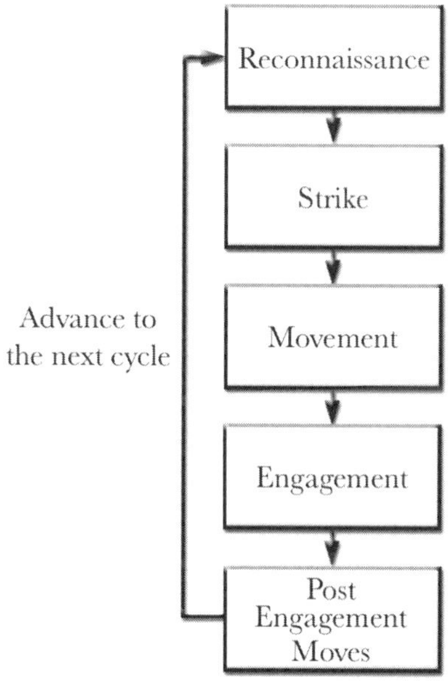

Figure 4.6: FASTHEX Game Cycle Sequence

The Decision Problem

With the epitomising principle in mind and the constraints on gaming imposed by the FASTHEX model, we examine the decisions open to each commander with particular emphasis on the Blue commander. Great care has been taken to ensure that decisions required in actual combat had their equivalent

[2]FASTHEX uses a hexagonal game board much like IDAHEX. For these games, each hexagon is 7.5 km from face to face.

during the play of the wargame. This is made possible by the fact that players can override almost all computer-generated outcomes in the game. Therefore, realism can be imposed in those cases where the model obviously strays.

The commander makes operational and tactical decisions at each combat phase in the FASTHEX wargame cycle in keeping with his overall operational and tactical aims (Figure 4.6). The one exception of course is the engagement phase: the engagement is a consequence of the decisions taken by the Red and Blue commanders in the other phases.[3] Therefore at each cycle, the decision set in FASTHEX, $D(t)$, consists of 4 components, $D(t)=\{d_1(t),d_2(t),d_3(t),d_4(t)\}$, where t indicates the cycle and $d_i(t)$ is the decision taken at the ith phase in the FASTHEX model. The following is a description of the decisions taken at each phase:

1. RECONNAISSANCE, $d_1(t)$**:** The commander must decide which of the reconnaissance assets at his disposal to allocate. In most cases, this means deciding where to direct his sensors and how many to commit to the process (taking account of higher level assets such as satellite surveillance). It should be noted here that reconnaissance assets are used primarily to identify valid targets and main force concentrations.

2. STRIKE, $d_2(t)$**:** The strike decisions can be thought of as the allocation of deep fire assets. The engagement phase adjudicates the close battle, but the deep fire battle is handled separately.

[3]Engagement adjudication is done using look-up tables based on lower level computer modelling of the various types of engagement.

3. MOVEMENT, $d_3(t)$: The commander must decide which of his units to move and how far they are to move. He is constrained by terrain, the maximum speed of his units, and the degree to which the units are fit (in terms of damage inflicted) to accomplish movement. Units move at the lowest level of resolution: the battalion, battle group, or squadron.
4. POST-ENGAGEMENT MOVES, $d_4(t)$: After each engagement, the commander assesses the damage done to his units. If the units are not capable of continuing as an integral force, they can be withdrawn, consolidated with other forces, or both.

OPTIMAL CONTROL FORMULATION

Each of the decisions is taken from among a discrete set of the possible choices described above and therefore the set $\{D(t)\}$ consisting of all possible decisions at cycle t has cardinality equal to the product of the cardinality of the 4 phase-decision sets. The collection of decision sets at each cycle in the game is then referred to as the *decision stream* for that game and therefore, we denote

$$D_G = \{D(0), D(1), \cdots, D(m-1)\}$$

the decision stream for an m-cycle game. Clearly, the number of possible decision streams for even a simple wargame can easily become unmanageably large and thus we are burdened by the "curse of dimensionality."

The consequences of the commander's decisions at each cycle can be measured in several ways. As mentioned earlier, we have selected the status of friendly and enemy forces. Consequently, if we let $X(t) = [x_1(t), x_2(t), \cdots, x_b(t), y_1(t), y_2(t), \cdots, y_r(t)]^T$

be the vector of combat strength of the *b* friendly force components and the *r* enemy force components at cycle *t*, then the decision stream can be viewed as a *memoryless multistage decision process* as depicted in Figure 4.7. At each cycle, the commander wishes to select *D(t)* so that a performance function dependent upon the vector *X(t)* is optimised in some way.

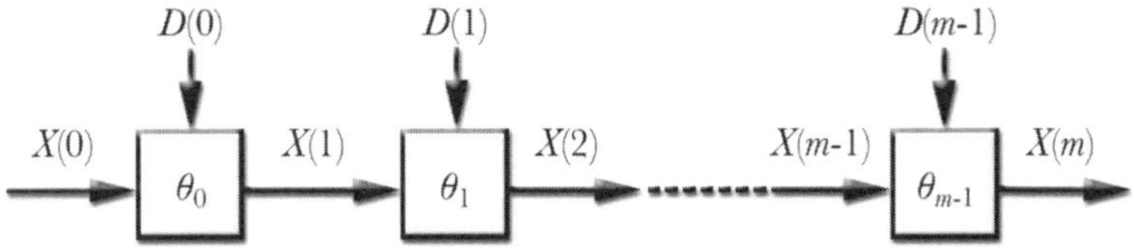

Figure 4.7: A Wargame as an Open Dynamical Process

In this formulation, θ_t is a transition function so that $X(t+1)=\theta_t[X(t),D(t)]$. The status of both Red and Blue forces in terms of combat strength in cycle *t*+1 depends upon the their combat status in cycle *t* and the decisions made in cycle *t*. The initial condition *X*(0) is the total combat strength of the friendly and enemy forces at the beginning of the campaign, and *X*(*m*) is the status of both at the termination of the campaign.

The commander's problem then is to select the decision stream D_G that optimises the performance (i.e., utility) function:

$$P=\sum_{t=0}^{m-1} f_t[X(t),D(t)]+\Phi[X(m)].$$

Assuming that we are able to find a reasonable representation for $f_t(\cdot)$ and $\Phi[X(m)]$, finding the optimal decision stream is then an optimal control problem of the form discussed in Chapter 1 and Appendix 1. In this case, the decision variables *D*(*t*) at each cycle are the control variables and the *X*(*t*) are the state variables. In Chapter 1 and Appendix 1, we discussed under what conditions we might expect such a problem to have a unique solution. For further discussion of such solutions, see also Bryson and Ho [14].

THE TWO-SIDED GAME

The problem with a two-sided game is the development of a second performance function for the opposing side. This problem can easily be solved if we design P such that if Blue chooses to maximise P, then Red will choose to minimise P. In this formulation, we assume two decision streams, one for the friendly commander $B_G = \{B(0),B(1),\ldots,B(m-1)\}$ and one for the enemy commander $R_G = \{R(0),R(1),\ldots,R(m-1)\}$. Our assumption implies that the game is zero-sum, that is, any increase in P for the friendly force results in an equal decrease for the enemy force and vice versa. Consequently, we wish to

$$\underset{B(t)\in B_G}{\text{maximise}}\ \underset{R(t)\in R_G}{\text{minimise}}\,(P),$$

subject to the transition constraint:

$$X(t+1)=\theta_t\left[X(t),B(t),R(t)\right],\ 0\le t\le m\text{-}1\,.$$

Solutions to problems of this type are fairly complex for all but very trivial examples. For example, see Hillestad [15] and Berkovitz and Dresher [16]. However, the objective here is not to solve the wargame using one of these techniques, but rather to use the two-sided memoryless multistage optimal control formulation as a convenient construct for a formal statement of the problem.

Games with Equivalent Decision Streams

Given the complexity of the two-sided optimal control construct for the wargames, we forego attempts to apply any closed form solution and rely instead on the replication of instances of the game from each of the scenarios and for the several cases to be examined. Even this however can be extremely time consuming and therefore we wish to examine those alternative cases in which the decision streams are essentially different. That is, if the introduction of alternative weapon systems in a game does not significantly alter the decision stream, D_G, then the two games are considered equivalent. In this way, the results of one game can be rerun under the conditions of the second. Although the results may vary with each, the decision stream is taken to be constant.

A simple example will serve to illustrate the process. Consider a conflict in which Red and Blue commanders have only artillery and tanks with which to conduct a two-cycle campaign, and we focus on the use of artillery resources. Our state vector then is $X(t)=[x_1(t),x_2(t),y_1(t),y_2(t)]^T$ where:

> $x_1(t), y_1(t)=$ the number of BLUE and RED artillery pieces respectively, and
>
> $x_2(t), y_2(t)=$ the number of BLUE and RED tanks respectively.

In both cycles, we assume that the decision on both sides is the allocation of artillery fires. Consequently, $B(t)=[b_1(t),b_2(t)]^T$ is the Blue commander's decision at cycle t $(t=0,1)$ and $R(t)=[r_1(t),r_2(t)]^T$ is the Red commander's decision, where $b_1(t),r_1(t)$ is the fraction of Blue/Red artillery allocated to attack Red/Blue artillery, $b_2(t)=1-b_1(t)$ is the fraction of Blue artillery to be allocated against Red tanks, and $r_2(t)=1-r_1(t)$ is the fraction of Red artillery to be allocated against Blue tanks. For simplicity, we restrict the domain to 0, .5, and 1. The two-cycle game thus described is illustrated in Figure 4.8:

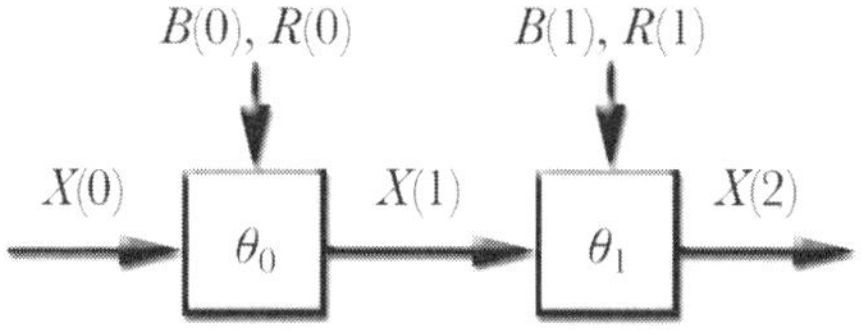

Figure 4.8: An Example Two-Cycle Game

The transition function, θ_t , is simply the combat adjudication model. If we assume a simple Lanchester differential model and let i = 1 for Blue and i = 2 for Red,

$$\theta_t[X(t),B(t),R(t)]=[\alpha_{21}r_1(t)y_1(t),\alpha_{22}r_2(t)y_1(t),\alpha_{11}b_1(t)x_1(t),\alpha_{12}b_2(t)x_1(t)]^T,$$

where:

$0 \leq \alpha_{i1} \leq 1$ is the effect of Blue/Red artillery against Red/Blue artillery; and

$0 \leq \alpha_{i2} \leq 1$ is the effect of Blue/Red artillery against Red/Blue tanks.

The α_{ij} s can be thought of as single shot kill probabilities (SSKPs) and $b_j(t)x_1(t)$ and $r_j(t)y_1(t)$ represent the number of Blue/Red artillery allocated to Red/Blue artillery and tanks. Therefore the transition equations become:

$$\begin{aligned}
x_1(t+1) &= x_1(t) - \alpha_{21}r_1(t)y_1(t) \\
x_2(t+1) &= x_2(t) - \alpha_{22}r_2(t)y_1(t) \\
y_1(t+1) &= y_1(t) - \alpha_{11}b_1(t)x_1(t) \\
y_2(t+1) &= y_2(t) - \alpha_{12}b_2(t)x_1(t)
\end{aligned}$$

We must also have that $\alpha_{2j} r_j(t) y_1(t) \leq x_j(t)$ and similarly, $\alpha_{1j} b_j(t) x_1(t) \leq y_j(t)$. In other words, the number of kills cannot exceed the number of target weapon systems available.

What remains to be defined is a utility function that is some measure of how well both sides accomplish their mission. For this simple problem, we assume that both sides wish to maximise the number of tanks available at the end of the second cycle. Their reasoning is that as the opposing sides come into closer contact, tanks are more effective than artillery. A utility function that does this is:

$$P = \sum_{t=0}^{1} \left[x_2(t) - y_2(t) \right] + 0.9 \left[x_2(2) - y_2(2) \right] + 0.1 \left[x_1(2) - y_1(2) \right].$$

The Blue commander therefore wishes to maximise *P* and the Red commander wishes to minimise *P*. Note that the weights, 0.9 and 0.1, reflect the relative importance assigned to tanks and artillery by the two commanders. Of course, this objective function is not unique. There are several other possibilities.

An interesting problem arises when the opposing commanders do not agree on a common objective. That is, what happens when the game is *not* zero-sum? In this case, we would need to evaluate a separate objective function for each side. In the example above, this would correspond to having a utility *P*(Blue) and a utility *P*(Red) that might correspond (for example) to different weightings for the value of tanks versus artillery due to different perceptions of the endstate (Red may simply wish to survive with a roughly balanced force, for example). We would then require some higher level measure of what this set of outcomes implies. In some cases (such as Operations Other Than War), we might be seeking to maximise the utility of both sides (i.e., a win-win situation rather than the win-lose assumption we made above).

For the example we have considered so far, we can now analyse an equivalent decision streams case. The allocation of artillery to enemy tanks and artillery over the two cycles is referred to as the allocation strategy. Figure 4.9 lists all the possible strategies by describing the game for the Blue commander in extensive form. Note that the diagram reflects only the allocation of Blue artillery to Red tanks, $b_2(t)$.

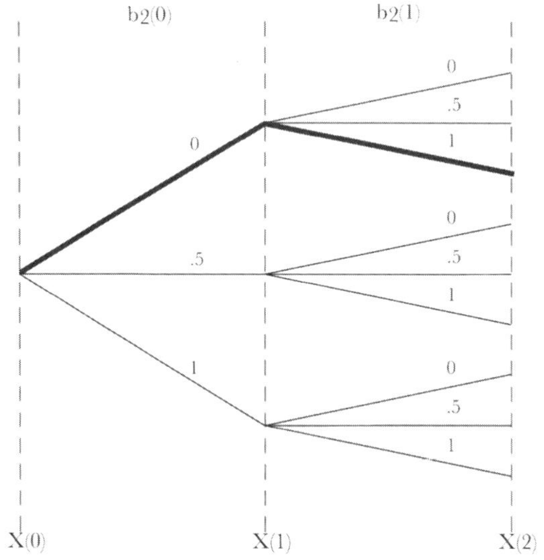

Figure 4.9: BLUE Commander's Allocation Strategy

The Blue commander reasons that during the first cycle, because the opposing forces are not in direct contact, the greatest threat to his forces is the enemy artillery. Therefore, he allocates all of his artillery against Red artillery. In the second cycle, as the forces begin to close, he sees Red tanks as the more serious threat and therefore allocates all of his resources against Red tanks. Therefore, his decision stream (allocation strategy) is:

$$B_G = \{B(0), B(1)\} = \{[1,0],[0,1]\}.$$

This is reflected in Figure 4.9 by the bold path, and corresponds to the *bang-bang* solution discussed in Appendix 1, where only extreme settings of the control variable are used. (This solution would result from a Lanchester square law Blue/Red interaction for example, which would give rise to a linear Hamiltonian, as explained in Appendix 1.)

Now, assume that a new game is to be played that differs from the current game only in that the Blue force has been augmented by more tanks. If the addition of these tanks does not alter the decision stream, then we consider them equivalent and the game may be replayed with the same decisions on both sides and the outcomes (value of P) compared.

Decision Uncertainty

ASTOR's primary function is to contribute to tactical situation assessment by observing the battlefield, detecting and identifying enemy units, and reporting on its findings. Consequently, a metric designed to measure how well situation assessment is accomplished in all cases tested was seen as useful to this study of the ASTOR sensor system. Such a metric allows us to measure *the degree of confidence the commander has that he possesses an accurate picture of the battlefield in his area of interest.* We would expect that the greater his knowledge about the location, size, and composition of the enemy force, the greater his confidence in making decisions concerning the allocation of his weapons and the movement of his forces. We also recognise that information of this type is not *all* he would require. Information concerning enemy intent gleaned from COMINT, SIGINT, and known enemy fighting doctrine would also assist in completing the picture.

We developed such a metric that reflects the amount of knowledge the commander has concerning the enemy forces arrayed against him in his area of interest. The measure is a function of the size, diversity, and effectiveness of the sensor suite, and the effectiveness of the command and control system used to process the reported sensor observations. The detailed development of this metric is covered next.

Probability Distribution

We begin by letting the vector U represent the competing hypotheses that any number of enemy units are arrayed against the friendly commander at time cycle t so that $U = \{0,1,2,...,n\}$. Given the level of resolution for the ASTOR games, a unit was taken to be a battalion. We omit the cycle index, t, for now focusing instead on analysis within a timestep. The term *arrayed against* is taken to indicate the units located on the battlefield in some area of interest to the friendly commander. This may mean along some avenue of approach in a defensive operation or blocking a route of advance in an offensive operation. Figure 4.10 depicts a notional defensive campaign situation.

We assume that the friendly commander knows the number of enemy units that might be brought to bear against him during the campaign. That is, we assume that he knows n. This is a reasonable assumption in that it is highly likely that the Intelligence Preparation of the Battlefield (IPB) would yield this information. What is unknown is the tactical deployment of the units at each timestep. Tactical situation assessment then is taken to be the process of estimating the enemy's tactical deployment at time t and the effectiveness of this estimate is the degree of uncertainty associated with his current state of knowledge.

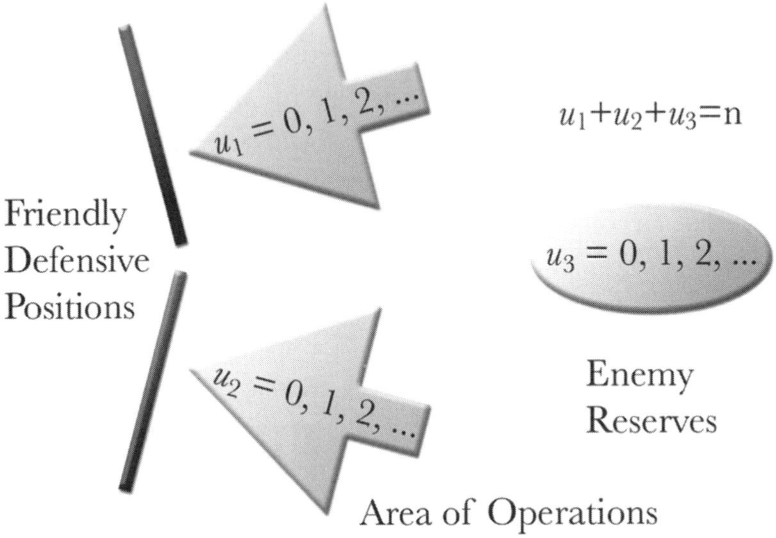

Figure 4.10: BLUE Commander's Situation Assessment Problem

BAYESIAN DECISIONMAKING

We begin by analysing the intelligence gathering process at each timestep. We first assume that a Bayesian update methodology for tactical situation assessment is appropriate within a wargame cycle, but not between wargame cycles, given the assumptions concerning the Markov properties (i.e., lack of memory) of the FASTHEX game with 2-hour timesteps.[4] Consequently, the process described here is repeated prior to each decision to commit forces.

1. INPUT DISTRIBUTION: The friendly commander may or may not have some idea of the likely disposition of enemy units. If he does, we may describe it using an empirical distribution. However, for this analysis, we assume that the friendly commander is completely igno-

[4]We later exploit Bayesian updating by assuming multiple sensor sweeps within a single decision cycle.

rant of the enemy commander's intentions. This provides us with a worst case situation, corresponding to the assumed lack of memory between timesteps. We let $P(U = u)$ represent the probability that the enemy commander will commit u of his n units in a specified area of interest in the area of operations AO (avenue of approach in Figure 4.10). Assuming that the enemy commander is equally likely to deploy any number of units in the area of interest, we have that $P(U = u) = 1/n+1$. The friendly commander hopes to refine this distribution using his sensor assets.

2. THE SENSOR MODEL: We next let $V = \{0,1,2,\ldots,n\}$ represent the number of units detected by the sensor assets allocated to the area of interest.[5] Therefore, $P(V = v)$ is the probability that the sensors will detect v of the enemy units arrayed against the friendly forces. However, this number is conditioned on the number of units in the area of interest deployed by the enemy commander. Consequently, we focus on the conditional probability, $P(V = v \mid U = u)$. For simplicity, we assume a single sensor is cued to search the specified area of interest.[6] We further assume that the sensor is capable of detecting a unit in the area of interest with probability q and that there are no false detections from the sensor or elsewhere.[7] Consequently, the conditional probability distribution, $P(V = v \mid U = u)$, is the binomial distribution:

[5]By "detect" we mean that sufficient information is provided to allow the unit to be targeted by a weapon.

[6]This assumption can be relaxed to allow for the characterisation of a multisensor suite, provided that the sensors are independent.

$$P(V = v \mid U = u) = \begin{cases} \binom{u}{v} q^{v} (1-q)^{u-v} & \text{for } v \le u \\ 0 \text{ otherwise.} \end{cases}.$$

3. SENSOR OPERATIONS: Our objective is to clarify the enemy force deployment picture based on the sensor observations by refining the friendly commander's initial and subsequent probability distributions on *U*. That is, we wish to calculate $P(U = u \mid V = v_d)$, where v_d is the number of detections reported in the cycle, and thus assess the impact of the evidence provided by the sensor on our estimate of the number of enemy units arrayed against the friendly forces in the area of interest. Operationally, we assume that the sensor sweeps the area of interest once in a cycle. As a detection occurs, it is immediately reported so that there are v_d+1 reports from the sensor per cycle. The additional report accounts for the fact that a report of 0 detections is sent initially. Since it is impossible to control the time when detections occur within a FASTHEX game cycle, we assume a uniform distribution of reports. That is, a report of no detections occurs at time $t/(v_d+1)$, a report of one occurs at $2t/(v_d+1)$, etc. The estimate is refined at every subinterval using Bayes' formula as follows:

$$P(U = u \mid V = v) = \frac{P(U = u \mid V = v-1) P(V = v \mid U = u)}{\sum_{i=0}^{n} P(U = i \mid V = v-1) P(V = v \mid U = i)} \quad (2)$$

[7]We later relax this assumption by allowing for the possibility that the sensor detections/identifications are false, that the command and control system used to transmit the sensor information may report a false detection/identification as real, and that the intelligence processing centre may interpret a false detection/identification as real.

In this formulation, $P(U=u \mid V=v-1)$ is the prior probability, $P(V=v \mid U=u)$ is the knowledge contributed by the latest report (the probability that one more unit is detected), and $P(U=u \mid V=v)$ is the posterior probability on U given the last report. Note that $P(U=u \mid V=-1)=P(U=u)=1/(n+1)$; that is, the prior distribution before sensors are deployed is flat, as described above. This process is repeated for $v = 0,1,\ldots,v_d$. Making the appropriate substitutions in (2), we get:

$$P(U=u \mid V=v)=\frac{P(U=u \mid V=v-1)\binom{u}{v}q^{v}(1-q)^{u-v}}{\sum_{i=0}^{n} P(U=i \mid V=v-1)\binom{i}{v}q^{v}(1-q)^{i-v}}$$
$$=\frac{P(U=u \mid V=v-1)\binom{u}{v}(1-q)^{u}}{\sum_{i-v}^{n} P(U=i \mid V=v-1)\binom{i}{v}(1-q)^{i}}, \qquad (3)$$

where $v = 0,1,\ldots,v_d$ is the number of units detected by the sensor and $u \geq v$ at each iteration. Figure 4.11 depicts the process diagrammatically. Note the difference between no sensor sweep in progress and a report of no detections. The former is depicted by a flat probability distribution on U whereas the latter is a refinement to the flat distribution.

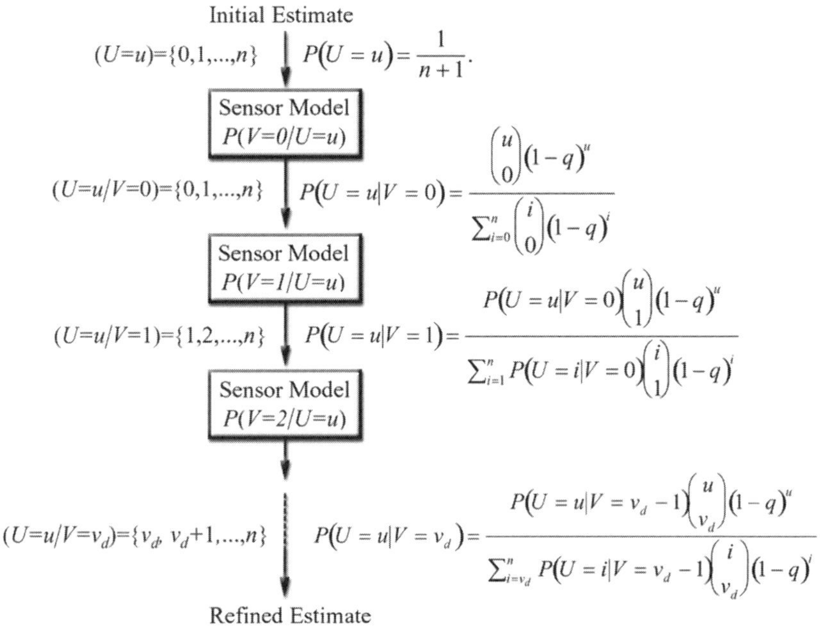

Figure 4.11: Developing a Refined Estimate

A SIMPLE EXAMPLE

The following illustrates the process. Table 4.1 summarises the results of a simple situation in which three units are known to be available to the enemy commander. The sensor system has a probability of detection/identification of $q = 0.8$. The entries in the rows are the refined probabilities from 0, 1, 2, and 3 detections. The first row is the a priori probability assessment on U assuming initial total ignorance. Figure 4.12 depicts the results graphically.

v	$P(U=0\mid V)$	$P(U=1\mid V)$	$P(U=2\mid V)$	$P(U=3\mid V)$
-	.2500	.2500	.2500	.2500
0	.8013	.1603	.0321	.0064
1	0	.9218	.0736	.0046
2	0	0	.9634	.0366
3	0	0	0	1

Table 4.1: Refined Probability Assessments: Example 1

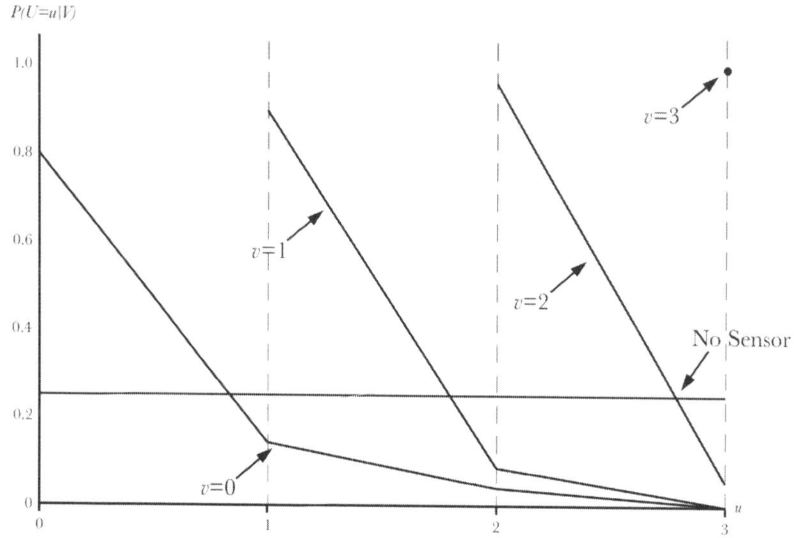

Figure 4.12: Refined Probability Assessments

MULTIPLE SWEEPS

We now refine the analysis to show the value that multiple sensor sweeps within the same cycle have on refining the probability estimates for *U*. Suppose that we assume that our sensors are capable of *k* sweeps of the area of interest within the commander's decision cycle. That is, the sensor can perform *k* sweeps of the area of interest before the enemy commander can move his units in any significant way. In each of these sweeps (i), v_{di} enemy units are detected. We further assume that the probability estimates are made sequentially, and that the sweep time is sufficiently small to allow for a single "end of sweep" report. Using Bayes formula we get:

$$P(U=u \mid V=v_{di}) = \frac{P(U=u \mid V=v_{d(i-1)})P(V=v_{di} \mid U=u)}{\sum_{j=0}^{n} P(U=j \mid V=v_{d(i-1)})P(V=v_{di} \mid U=j)} \tag{4}$$

where $i = 1,2,\ldots,k$, and $P(U=u \mid V=v_{d0}) = P(U=u) = 1/n+1$. Making the appropriate substitutions, (4) becomes:

$$P(U=u \mid V=v_{di}) = \frac{P(U=u \mid V=v_{d(i-1)})\binom{u}{v_{di}}(1-q)^{u}}{\sum_{j=v_{di}}^{n} P(U=j \mid V=v_{d(i-1)})\binom{j}{v_{di}}(1-q)^{j}}. \tag{5}$$

In general, Bayesian updating has a tendency to converge rather rapidly—especially in cases such as this where false detections/identifications are not allowed: that is, it is impossible to overstate the number of units actually present. The effect is that subsequent detections that report fewer units than the previous are totally ignored. To illustrate, consider a simple case in which $n = 3$ units. We assume that three sweeps were conducted resulting in three sequential detections using a sensor with probability of detection: $q = 0.8$. Table 4.2 summarises the results of applying equation (5) with $k = 3$. The number of units detected each time is listed in the table. The number of units in the area of interest is actually three and subsequent observations that two units were detected/identified are completely ignored.

Multiple Sweeps

i	v_{di}	$P(U=0 \mid V)$	$P(U=1 \mid V)$	$P(U=2 \mid V)$	$P(U=3 \mid V)$
0	-	0.250	0.250	0.250	0.250
1	2	0	0	0.625	0.375
2	3	0	0	0	1
3	2	0	0	0	1

Table 4.2: Multiple Sweeps Case 1

Now consider a second case with a somewhat different history as depicted in Table 4.3. In this case, four sweeps were conducted resulting in the sequential detections depicted in the Table. The detection of one unit persisted for three reports. Note the rapid convergence of $P(U=1 \mid V)$. However, the single detection of two units in Sweep 4 shifts the mode of the distribution to $U = 2$. Because we exclude false detections, all reports less than the current number detected will be ignored.

i	v_{di}	$P(U=0 \mid V)$	$P(U=1 \mid V)$	$P(U=2 \mid V)$	$P(U=3 \mid V)$
0	-	0.250	0.250	0.250	0.250
1	1	0	0.658	0.263	0.079
2	1	0	0.767	0.122	0.111
3	1	0	0.925	0.059	0.016
4	2	0	0	0.855	0.145

Table 4.3: Multiple Sweeps Case 2

FALSE TARGET DETECTIONS/ IDENTIFICATIONS

Up to this point, we have assumed that false targets were not present. It is possible to relax this assumption and recognise that targets can be misclassified in several ways: (1) the sensor system may be defective; (2) the command and control system used to transmit the detection to a central processing centre may have erroneously introduced a false target; (3) the processing centre equipment or personnel may have misinterpreted the data being received; and (4) the enemy may be actively engaging in deception activities (i.e., Information Operations). The way in which this is done is described in detail in reference [11], using a Poisson process to represent the "flow" of false targets through the sensing process.

KNOWLEDGE REPRESENTATION

It now remains to ascertain the degree of uncertainty existing in the mind of the friendly commander at the time he must take a decision on the employment of his forces. His current knowledge consists of two components: (1) the fact that his sensor suite detected a number of enemy units in his area of interest; and (2) the refined probability distribution over the possible number of enemy units that might be in his area of interest based on his most recent sensor report. The value of the first component depends upon whether false detections are possible. The second depends upon the number of enemy units detected and the reliability of the sensor system. The task is to develop a *knowledge metric* that incorporates these two components, thereby quantifying the likelihood that the commander has a true picture of the number of enemy units arrayed against him in his area of interest.

INFORMATION ENTROPY

We draw on information science to develop a knowledge metric that is a function of the average information present in the set of all possible uncertain events. This quantity is referred to as *information entropy*[8] and it measures the *amount of uncertainty* in a probability distribution.

The amount of information available from the known occurrence of the event, $U = u$, i.e. that u enemy units are indeed arrayed against the friendly force, is inversely proportional to the likelihood that the event will occur. An event that is very likely to occur provides little information when it does occur. On the other hand, an unlikely event provides considerable information when it occurs. Mathematically, we define information as follows:

$$I(U = u) = \ln \frac{1}{P(U = u)} = -\ln P(U = u).$$ [9]

If we now consider all of the events in the refined set $U \mid V = v_d$, we reason that each occurs with probability $P(U = u \mid V = v_d)$. Therefore, the information available from the occurrence of each event is:

$$I(U = u \mid V = v_d) = -\ln P(U = u \mid V = v_d),$$

and the expected information from the occurrence of each event is:

$$P(U = u \mid V = v_d)\, I(U = u \mid V = v_d) = -P(U = u \mid V = v_d) \ln P(U = u \mid V = v_d).$$

---

[8]The term *entropy* is used because the information entropy function is the same as that used in statistical mechanics for the thermodynamic quantity entropy. For a more complete discussion of entropy, see Blahut [17] and Zurek, ed. [18].

[9]In communication theory, the units of measurement are "bits" if base 2 logarithms are used and "nits" if natural logarithms are used (See Kullback [19] p. 7). For our purposes, we will assume a dimensionless quantity.

Consequently, the average amount of information in the probability distribution: $P(U \mid V = v_d)$ can be expressed as:

$$H[P(U \mid V = v_d)] = H(U \mid V = v_d) = -\sum_{i=0}^{n} P(U = i \mid V = v_d) \ln[P(U = i \mid V = v_d)].$$

The entropy quantity $H(U \mid V = v_d)$ is the *residual uncertainty* regarding U given that V is instantiated to v_d. The average uncertainty then is the sum of the residual uncertainties weighted by the probability distribution on the sensor detection/identifications:

$$H(U \mid V) = -\sum_{j=0}^{n} P(V = j) \sum_{i=0}^{n} P(U = i \mid V = j) \ln[P(U = i \mid V = j)].$$

Properties of Information Entropy

Information entropy has properties that make it ideal as a metric for measuring the commander's uncertainty prior to making a decision and for measuring the uncertainty in the entire campaign:

1. Maximum Entropy: The entropy function is maximised when the uncertainty in the distribution is greatest. Maximum uncertainty occurs when the friendly commander has no sensor assets to deploy. In this case, any number of units might be arrayed against him with equal probability. Mathematically we have that $P(U = u) = 1/n+1$. The entropy in this case is:

$$H(U) = -\sum_{i=0}^{n} \frac{1}{n+1} \ln \frac{1}{n+1} = \ln(n+1).$$

   Thus the maximum uncertainty in *P(U)* is $\ln(n+1)$. Note that as n grows larger, the entropy increases as we might expect; the more units available to the enemy commander, the less clear we are about their deployment in the absence of sensor outputs. In general, a probability distribution with a wide variance exhibits high entropy.

2. MINIMUM ENTROPY: The entropy function is minimised at 0. This occurs when $P(U = u_i) = 1.0$ and $P(U = u_j) = 0$ for all $j \neq i$. This represents total certainty or minimum uncertainty.

3. CAMPAIGN ENTROPY: The total campaign entropy, denoted $H(U_1, U_2, \ldots U_m)$, where m is the total number of game (campaign) cycles satisfies the relation:

$$H(U_1, U_2, \cdots U_m) \leq \sum_{i=1}^{m} H(U_i).$$

The equality condition holds when the process is memoryless as approximated in the FASTHEX games (for purposes of tactical situation assessment), when the situation being considered is rapidly changing across the timespan of the campaign.

COMBAT CYCLE KNOWLEDGE

For the wargames we were dealing with, we found it important to develop a metric that was capable of providing an ordinal ranking of the wargame cases across all scenarios in terms of the knowledge possessed by the commander prior to making a decision *at each cycle.* Although entropy is a convenient measure of decision uncertainty, making direct comparisons among the cases examined could be misleading. We need only recall that maximum entropy is $H(U) = \ln(n+1)$ to realise that the varying number of enemy units in the AO makes a direct comparison incorrect. In addition, it is incomplete because it addresses only the second knowledge component, namely the knowledge gained from the refined probability distribution. What is needed is a more comprehensive metric incorporating the *residual uncertainty* in the refined distribution, and the *detection information* gained by the sensor report. Having said this, in later applications of this idea to quantifying Information Dominance [6] and the benefits of a network-centric approach to collaboration and task prosecution [9], the simpler form of Residual Uncertainty (and hence, Residual Knowledge–which is a measure of uncertainty removed) has been found to be adequate.

For the wargaming application then, we let $K(U,V=v_d)$ represent the knowledge gained from detecting v_d enemy units when there are U enemy units in the area of interest. Symbolically we have:

$$K(U,V=v_d)=K(U\mid V=v_d)K(V=v_d)$$

where $K(U\mid V=v_d)$ is the knowledge associated with the residual uncertainty in the refined probability distribution given a sensor report of v_d units, and $K(V=v_d)$ is the knowledge gained by detecting/identifying v_d enemy units. If we can ensure that both $K(U\mid V=v_d)$ and $K(V=v_d)$ are confined to the interval [0,1], then we can think of $K(U,V=v_d)$ as a probability.[10] As such, it represents the likelihood that the commander has a complete picture of the battlefield at the time he makes a decision. This can be a very powerful statistic when correlated with the Force Loss Exchange Ratio, enemy attrition, and friendly survivability as discussed below.

1. RESIDUAL KNOWLEDGE: The maximum uncertainty in $P(U\mid V=v_d)$ is $\ln(n+1)$. Therefore, maximum *certainty*

[10] $K(U,V=v_d)$ satisfies the probability axioms (see Stark and Woods [13] p. 9 for instance) and therefore can be thought of as a subjective probability.

can be defined as $\ln(n+1) - H(U \mid V = v_d)$.[11] Normalising this quantity provides us with the following definition of residual knowledge:

$$K(U \mid V = v_d) = \frac{\ln(n+1) - H(U \mid V = v_d)}{\ln(n+1)}. \tag{6}$$

Residual knowledge is maximised when residual entropy is 0 and it is minimised when residual entropy is $\ln(n+1)$. In general, residual knowledge reflects the amount of uncertainty in the refined probability distribution.

2. DETECTION KNOWLEDGE: Given that v_d enemy units were detected, we are now concerned with the additional information this provides concerning the likelihood that there are actually v_d or more enemy units in the area. This is clearly a function of the reliability of the sensors and the command and control system used to process the sensor data it receives. Mathematically, we are interested in the information content for the event: $U \geq v_d \mid V = v_d$. That is, the information that will be provided from the detection reports this cycle, or the *prior* information content of the event, $V = v_d$. This is calculated to be:

$$I(U \geq v_d \mid V = v_d) = -\ln[P(U \geq v_d \mid V = v_d)] = -\ln\left[\sum_{i=v_d}^{n} P(U = i \mid V = v_d)\right]$$

If $v_d = 0$, we get no information because $P(U \geq 0) = 1$. However, if $v_d = n$, the information content is maximised at $-\ln\left[P(U = n \mid V = v_d)\right]$. This is due to the fact that $P(U \geq u \mid V = v_d)$ decreases monotonically with

[11]In general, the change in information resulting from detecting $V = v_d$ units is $\Delta I(U \mid V = v_d) = H(U) - H(U \mid V = v_d)$.

increasing u and therefore is smallest for $u = n$. This suggests the following definition for $K(V = v_d)$:

$$K(V = v_d) = \frac{\ln\left[\sum_{i=v_d}^{n} P(U = i \mid V = v_d - 1)\right]}{\ln\left[P(U = n \mid V = v_d - 1)\right]}.$$

(We use $v_d - 1$ to ensure that the denominator never goes to zero).

The total knowledge gained is then defined to be the product of residual and detection knowledge:

$$K(U,V = v_d) = \frac{\ln(n+1) - H(U \mid V = v_d)}{\ln(n+1)} \cdot \frac{\ln\left[\sum_{i=v_d}^{n} P(U = i \mid V = v_d - 1)\right]}{\ln\left[P(U = n \mid V = v_d - 1)\right]} \qquad (7)$$

We can apply equation 7 to the example depicted in Table 4.1. The first 5 columns of Table 4.4 repeat the information in Table 4.1 for convenience. The last two columns contain the entropy and knowledge figures based on the refined distributions at each iteration, and the intermediate values of V. Figure 4.13 depicts the results graphically.

v	$P(U=0 \mid V)$	$P(U=1 \mid V)$	$P(U=2 \mid V)$	$P(U=3 \mid V)$	$H(U \mid V)$	$K(U,V)$
-	.2500	.2500	.2500	.2500	1.3863	0
0	.8013	.1603	.0321	.0064	0.6130	0
1	0	.9218	.0736	.0046	0.2918	.1638
2	0	0	.9634	.0366	0.1570	.4434
3	0	0	0	1	0	1

Table 4.4: Total Knowledge: Example 1

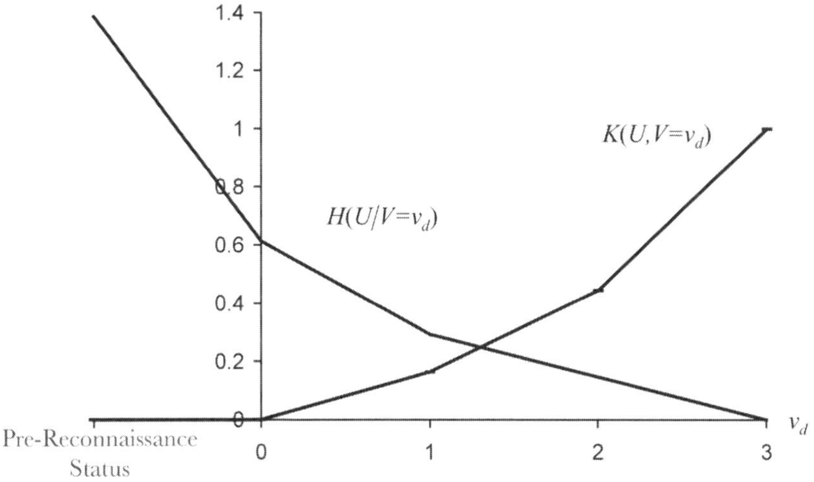

Figure 4.13: Knowledge and Entropy for Example 1

Campaign Knowledge

A similar formulation may now be used to calculate *campaign knowledge*, given that the FASTHEX games are taken to be memoryless processes for tactical situation assessment. Consider a campaign consisting of m cycles. At each cycle, t, v_{dt} enemy units are detected by the sensor. At each cycle, the number of possible enemy units arrayed against the friendly forces, n, is likely to be reduced as a result of combat during the cycle so that n_t is the total number of enemy forces that might be arrayed against the friendly forces in the area of interest. $H(U_t \mid V = v_{dt})$ then represents the residual uncertainty at each cycle and the total campaign uncertainty is expressed as:

$$H(U_1 \mid V = v_{d1}, U_2 \mid V = v_{d2}, \cdots, U_m \mid V = v_{dm}) = \sum_{t=1}^{m} H(U_t \mid V = v_{dt})$$
$$= -\sum_{t=1}^{m} \sum_{i=0}^{n_t} P(U_t = i \mid V = v_{dt}) \ln P(U_t = i \mid V = v_{dt}).$$

By analogue, residual knowledge for the entire campaign can be defined as:

$$K(U_1 \mid V = v_{d1}, U_2 \mid V = v_{d2}, \cdots, U_m \mid V = v_{dm}) = \frac{\sum_{t=1}^{m} [\ln(n_t + 1) - H(U_t \mid V = v_{dt})]}{\sum_{t=1}^{m} \ln(n_t + 1)}.$$

Detection knowledge can be calculated in a similar way. In this case, we are interested in the total information available from having detected/identified v_{dt} enemy units at each of the m cycles, t. To do this, we rely on the fact that the total information available from the occurrence of m independent events is the sum of the information available from the occurrence of each of them. Therefore we get that:

$$I(U_1 \geq v_{d1} \mid V = v_{d1} - 1, U_2 \geq v_{d2} \mid V = v_{d2} - 1, \cdots, U_m \geq v_{dm} \mid V = v_{dm} - 1)$$
$$= \sum_{t=1}^{m} I(U_t \geq v_{dt} \mid V = v_{dt} - 1).$$

Detection knowledge for the entire campaign can then be expressed as:

$$I(U_1 \geq v_{d1} \mid V = v_{d1} - 1, U_2 \geq v_{d2} \mid V = v_{d2} - 1, \cdots, U_m \geq v_{dm} \mid V = v_{dm} - 1)$$
$$= \frac{\sum_{t=1}^{m} \ln\left[\sum_{i=v_{dt}}^{n_t} P(U_t = i \mid V = v_{dt} - 1)\right]}{\sum_{t=1}^{m} \ln\left[P(U_t = n_t \mid V = v_{dt} - 1)\right]}$$

and total campaign knowledge is:

$$K(U_1, V = v_{d1}, U_2, V = v_{d2}, \cdots, U_m, V = v_{dm})$$
$$= \frac{\sum_{t=1}^{m} \left[\ln(n_t + 1) - H(U_t \mid V = v_{dt})\right]}{\sum_{t=1}^{m} \ln(n_t + 1)} \cdot \frac{\sum_{t=1}^{m} \ln\left[\sum_{i=v_{dt}}^{n_t} P(U_t = i \mid V = v_{dt} - 1)\right]}{\sum_{t=1}^{m} \ln\left[P(U_t = n_t \mid V = v_{dt} - 1)\right]}.$$

An Example

Consider the example summarised in Table 4.5. The campaign consists of 5 cycles. At each cycle t, the detection probability q_t, the number of units detected v_{dt} and the maximum size n_t of the enemy force is given. The last three columns depict the residual uncertainty in the refined probability distribution, the information available from the detection of v_{dt} enemy units, and the "probability" that the commander has an accurate picture of the number of enemy units in his area of interest. The last row reflects his total campaign knowledge.

t	q_t	v_{dt}	n_t	$K(U_t \mid V = v_{dt})$	$K(U_t = v_{dt})$	$K(U_t, V = v_{dt})$
1	.6	2	5	0.6122	0.2263	0.1385
2	.4	3	5	0.3950	0.3868	0.1528
3	.9	3	4	1.0000	0.5693	0.5693
4	.9	0	3	0.5572	0	0
5	.3	2	3	0.5166	0.5000	0.2583
Total Campaign Knowledge						0.2092

Table 4.5: Total Knowledge

The detection probability q_t is assumed to change as a function of time t to reflect the changing sensor mix. The maximum size of the enemy force n_t reduces over time to reflect attrition. The fluctuation in numbers detected v_{dt} (including a complete lack of detections during one time cycle) leads to a reduced value of overall campaign knowledge. This is scaled to vary between 0 and 1, with 0 representing complete ignorance, and 1 representing complete knowledge of the number of enemy units in the commander's area of interest at every stage of the campaign.

THE EFFECTS OF KNOWLEDGE

We stated at the outset that it was desirable to assess the effects of increased knowledge on the outcome of the campaign. One way to do this is to compare $K(U,V = v_d)$ to the Force Loss Exchange Ratio (FLER) and the friendly and enemy combat attrition. Comparisons with the FLER measure how knowledge influences the relative losses in combat. Comparisons with friendly and enemy attrition measure the degree to which knowledge enhances the survivability of friendly forces and the destruction of the enemy. Statistically, we have shown a positive linear relationship when the FLER or enemy attrition is compared to knowledge, and a negative linear relationship when friendly casualties are compared to knowledge [11]. The results also showed that the approach and structure (the use of the epitomising approach, and open gaming) adopted in the wargames was able to produce a set of coherent and quantified alternatives, which formed the basis of the balance of investment analysis. As a consequence of this analysis, information could be weighed in the same scale as weapon effects, and the benefit clearly demonstrated.

Figure 4.14 shows the relation between campaign level knowledge and attrition of enemy forces, as assessed using the FASTHEX wargaming experiments, giving a correlation value of 0.8.

Figure 4.15 shows the effect of an increase in campaign level knowledge on own force casualties, again as a result of the FASTHEX wargaming experiments, giving a correlation value of −0.4.

Combat Cycle Knowledge

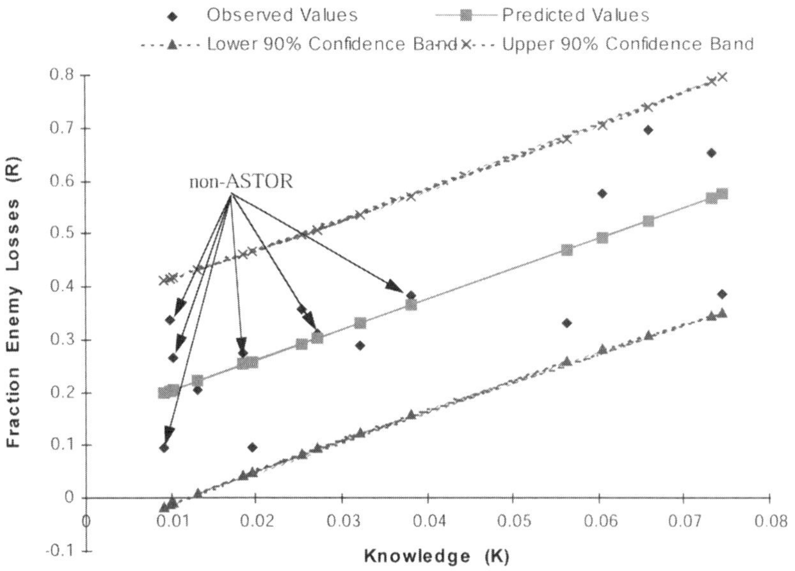

Figure 4.14: Experimental Assessment of Campaign Level Knowledge and Attrition of Enemy Forces

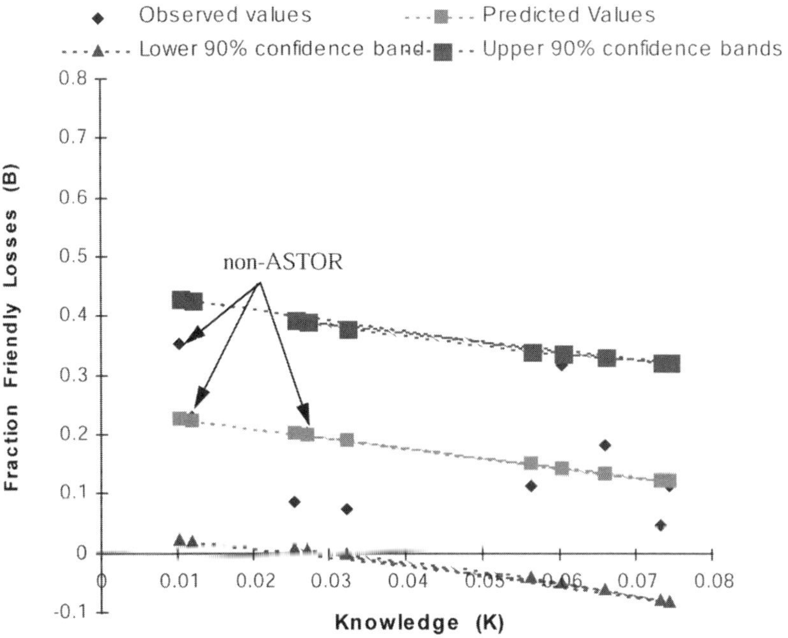

Figure 4.15: Experimental Assessment of the Effect of Campaign Level Knowledge on Own Force Casualties

As discussed in [11] from the experimental results, there appears to be a point where the knowledge available to the commander exceeds his capacity to act on it, either to generate more enemy losses or to prevent further friendly casualties, and we need to represent this effect.

Quantifying the Benefit of Collaboration Across an Information Network

In further exploitation of these ideas, Perry [9] has shown how this approach can be used as a basis for quantifying the benefit of collaborating across an information network. An example of the approach is described below, as used in recent work by Dstl in the UK in the context of modelling a time-critical operation. Full details of the general approach and other areas of application are in [9].

We assume we have a network of command and control nodes that are involved in coordinating a time-critical operation. Each of these nodes has a number of information processing tasks to perform. If:

$$\frac{1}{\lambda_i}$$

is the mean time for node i to complete all of its tasks, we assume that this completion time is distributed exponentially (an exponential distribution is used to model the time between events or how long it takes to complete a task), so that if $f_i(t)$ is the probability of completing all tasks at node i by time t, then:

$$f_i(t) = \lambda_i e^{-\lambda_i t}.$$

In general, there will be a number of parallel and sequential nodes in the network sustaining the operation. Let this total number be τ. In the simplest case, there is a critical path consisting of ρ nodes where ρ is a subset of τ, as shown in Figure 4.16.

Figure 4.16: The Critical Path

We define the total latency of the path as the sum of the delays (latencies) at each of the nodes, plus the time, defined as t_m, required to move a terminal attack system (such as an aircraft) to the terminal attack area. In this sequential case, we thus have that the total expected latency T is the sum of the expected latencies at each node on the critical path, plus the time t_m:

$$T = \sum_{i=1}^{\rho} \frac{1}{\lambda_i} + t_m .$$

If there are sequential and parallel nodes on the critical path, these can be dealt with in the way shown by the example below:

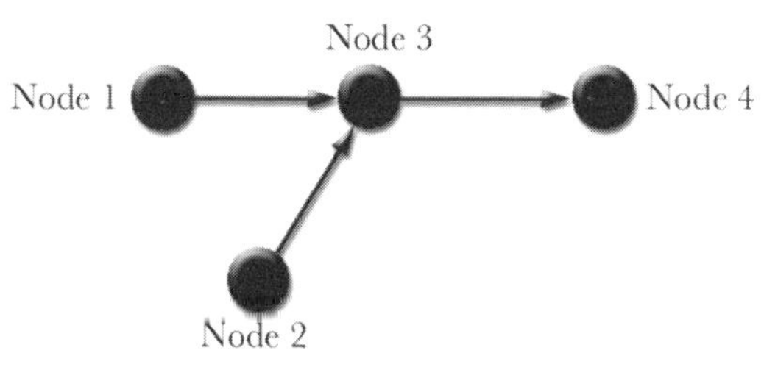

Figure 4.17: Parallel Nodes on the Critical Path

In this example:

$$T = \max(\frac{1}{\lambda_1}, \frac{1}{\lambda_2}) + \frac{1}{\lambda_3} + \frac{1}{\lambda_4} + t_m .$$

Returning now to the case of a serial set of nodes that constitute the critical path, for each such node i on the critical path define the *indegree* d_i to be the number of command and control (C2) network edges having i as a terminal link.

For each node j in the C2 network, we assume (based on our earlier discussion of information entropy and knowledge) that the amount of knowledge available at node j concerning its ability to process the information and provide quality collaboration is a function of the uncertainty in the distribution of information processing time $f_j(t)$ at node j. Thus the more we know about node j processes, the better the quality of collaboration with node j.

Let $H_j(t)$ be the Shannon entropy of the function $f_j(t)$. Then $H_j(t)$ is a measure of this (residual) uncertainty defined in terms of a lack of knowledge. By definition of the Shannon entropy, we have:

$$H_j(t) = -\int_0^\infty \ln(\lambda_j e^{-\lambda_j t}) \lambda_j e^{-\lambda_j t} dt$$

Since the differential of $(xe^x - e^x)$ is xe^x

it follows that $H_j(t) = \ln(\frac{e}{\lambda_j})$

If $\lambda_j$min corresponds to a minimum rate of task completions at node j, then:

$$\frac{1}{\lambda_j \min}$$

corresponds to a maximum expected time to complete all tasks at node j. In order to provide a normalised value of the knowledge $K_j(t)$ available at node j in terms of the Shannon entropy, we define this as:

$$
\begin{aligned}
K_j(t) &= \ln\left(\frac{e}{\lambda_j \min}\right) - \ln\left(\frac{e}{\lambda_j}\right) \\
&= \ln\left(\frac{\lambda_j}{\lambda_j \min}\right) \quad \text{if} \lambda_j \min \leq \lambda_j \leq e\lambda_j \min \\
&= 0 \text{ if } \lambda_j < \lambda_j \min \\
&= 1 \text{ if } \lambda_j > \text{e}\lambda_j \min
\end{aligned}
$$

Suppose now that node i is on the critical path, and node j is another network node connected to node i. Let c_{ij} represent the quality of collaboration obtained by including node j. If this is high, reference [9] assumes $K_j(t)$ will be close to 1. The effective latency at node i is thus assumed to be reduced by the factor $(1 - K_j(t))^{\omega_j}$ due to the effect of this high quality of collaboration. The factor ω_j is assumed to be 1 if j is one of the nodes directly involved in the time-critical operation (but not on the critical path). It is assumed to be 0.5 if node j is one of the other network nodes, to reflect a lower level of collaboration with these nodes.

It is important to note that the actual latency may not be reduced by this collaboration, but the ability to use the time more wisely through collaboration (to fill in missing parts of the operational picture that are available from other nodes, etc.) has an impact that can be expressed *equivalently* in terms of latency reduction. The use of such time more wisely implies a good knowledge of expected time to complete tasks that can provide such information. This is similar to the use of entropy and knowledge in the FASTHEX wargames where increased knowledge led to better awareness of the layout of the enemy force, and hence to wiser use of the commander's own forces. Such wiser use, due to information superiority, could be quantified in that case by an equivalent (linear) improvement in the number of enemy units destroyed, or a reduction in own force casualties.

The balance to be struck is that between such enhanced collaboration and the effects of information overload due to increasing network complexity (which we assess separately as a function of the number of elements of the network involved in the task).

The total (equivalent) reduction in latency at node *i* due to collaboration with the network nodes connected to node *i* is then given by:

$$\begin{aligned} c_i &= \prod_{j=1}^{d_i} c_{ij} \\ &= \prod_{j=1}^{d_i} (1 - K_j(t))^{\omega_j} \end{aligned}$$

Thus the total effective latency along the critical path, accounting for the positive effects of collaboration, is given by:

$$\begin{aligned} T_c &= \sum_{i=1}^{\rho} \frac{c_i}{\lambda_i} + t_m \\ &= \{\sum_{i=1}^{\rho} \frac{1}{\lambda_i} \prod_{j=1}^{d_i} (1 - K_j(t))^{\omega_j}\} + t_m \end{aligned}$$

We noted in the experimental data from the ASTOR study discussed earlier that we need to represent the effect of information *saturation*. Reference [9] also includes a *complexity penalty* to account for the fact that taking account of additional network connectivity leads to such information overload effects. This is the negative effect of collaboration. It leads to an increase in effective latency on the critical path. Following [9], we define C to be the total number of network connections accessed by nodes on the critical path. For each node i on the critical path, this is the indegree d_i. Thus:

$$C = \sum_{i=1}^{\rho} d_i \; .$$

The value of C is then a measure of the complexity of the network. We assume that the complexity effect associated with a particular value of C follows a nonlinear S-shaped curve as shown below.

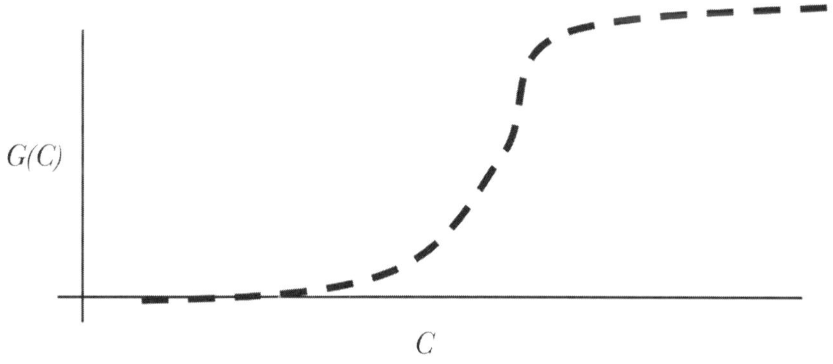

Figure 4.18: The Logistics S-Shaped Curve

The equation used to describe this effect is a Logistics equation:

$$g(C) = \frac{e^{a+bC}}{1+e^{a+bC}} \, .$$

The penalty for information overload is then defined as:

$$\frac{1}{1-g(C)} \, .$$

The total effective latency, taking account of both the positive and negative effects of C2 network collaboration, is then:

$$T_{c,C} = \frac{T_c - t_m}{1 - g(C)} + t_m \, .$$

Network-Centric Benefit

This network-enabled approach thus allows us to compute the distribution of the response time of the system as a function of the network assumptions. As we increase the collaboration throughout the network in going from platform-centric to network-centric to futuristic network-centric (to use the RAND categories [9]), so the positive effects of enhanced collaboration have to balance off against the downside effects of information overload and increasing network complexity. Going back to the discussion in Chapter 2 on the Conceptual Framework of Complexity, we can call this overall assessed performance of the network the *plecticity*[12] of the network, since it characterises the combined positive and negative effects of network complexity and collaboration.

Stochastic Networks and Network Vulnerability

So far, we have shown how it is possible to calculate the positive and negative effects of network complexity and collaboration based on the use of Shannon entropy as a measure of (lack of) knowledge. We can extend this model potentially in a number of ways. The length of the critical path, for example, if the network is adapting over the course of a series of tasks, will be a stochastic variable. We would expect,

[12] A term proposed by Perry (RAND Corp.) - personal communication.

from the theory we have considered so far that the size of the network (i.e., the number of nodes on the critical path) should be sampled from a power-law distribution of network size. The exponent of this power-law is then a characteristic measure of the ability of the nodes in the network to form and reform dynamically over time. Similarly, we can consider the indegree of a node on the network to be a stochastic variable. If the indegree of a node is sampled from a power-law distribution of the number of links, then the network is said to be "scale-free" [20]. This corresponds to a network with a small number of nodes with very rich connections, and many nodes with sparse connections. (The Internet is an example of a scale-free network.) Conversely, if the indegree of a node is sampled from a normal distribution of the number of links, then the network is of "random" type. Characterising networks in this way allows us to investigate the vulnerability of such networks to attack, as discussed in [20].

REFERENCES

1 MOFFAT J (2002). *Command and Control in the Information Age: Representing its Impact.* The Stationery Office. London, UK.

2 HORNE G E and LEONARDI M (2001). *Manoeuver Warfare Science 2001.* Marine Corps Combat Development Command. Quantico, VA, USA.

3 LAUREN M (2002). "Firepower Concentration in Cellular Automaton Combat Models – An Alternative to Lanchester." *J Opl Res. Soc* 53 Issue 6, pp. 672-679.

4 www.cna.org/isaac (Aug 1, 2003)

5 GOLDENFELD N (1992). *Lectures on Phase Transitions and the Renormalisation Group.* Addison-Wesley. MA, USA.

6 DARILEK R, PERRY W et al (2001). *Measures of Effectiveness for the Information Age Army.* RAND. Santa Monica, CA, USA.

7 TURCOTTE D L (1997). *Fractals and Chaos in Geology and Geophysics.* 2nd Edn. Cambridge University Press. Cambridge, UK.

8 http://fafnir.phyast.pitt.edu/myjava/perc/percTesT.html (February 1, 2003)

9 PERRY W, BUTTON R W et al (2002). *Measures of Effectiveness for the Information-Age Navy: The Effects of Network-Centric Operations on Combat Outcomes.* RAND. Santa Monica, CA, USA.

10 ROSKE V (2002). "Opening Up Military Analysis: Exploring Beyond the Boundaries." *Phalanx.* USA Military Operations Research Society. 35 No 2.

11 PERRY W and MOFFAT J (1997). "Measuring the Effects of Knowledge in Military Campaigns." *J Opl Res.* Soc 48. pp. 965-972.

12 BOWEN K C (1978). *Research Games – An Approach to the Study of Decision Processes.* Taylor and Francis, UK.

13 STARK H and WOODS J W (1986). *Probability, Random Processes and Estimation Theory for Engineers.* Prentice Hall, USA.

14 BRYSON A E and HO Y C (1975). *Applied Optimal Control.* Hemisphere Publishing, USA.

15 HILLESTAD R (1986). *SAGE: An Algorithm for the Allocation of Resources in a Conflict Model.* RAND working draft.

16 BERKOVITZ D and DRESHER M (1959). *A Game-theory Analysis of Tactical Air War.* Operations Research, 17. pp. 599-620.

17 BLAHUT R E (1987). *Principles and Practice of Information Theory.* Addison-Wesley. MA, USA.

18 ZUREK W H ed. (1990). *Complexity, Entropy and the Physics of Information.* Vol III, Santa Fe Institute Studies in the Sciences of Complexity Series. Addison Wesley. USA.

19 KULLBACK S (1968). *Information Theory and Statistics.* Dover. New York, USA.

20 COHEN D (2002). "All the World's a Net." *New Scientist.* 13 April 2002. pp. 24-29.

CHAPTER 5

AN EXTENDED EXAMPLE OF THE DYNAMICS OF LOCAL COLLABORATION AND CLUSTERING, AND SOME FINAL THOUGHTS[1]

Towards the end of Chapter 4, we discussed the way in which a particular network could be analysed using what we called the *plecticity* of a network, which includes both the positive effects of collaboration and the downside effects of information overload. This assumes a network of a particular size and configuration, and thus raises

[1] The contribution of Dr. Susan Witty, Dstl, to this chapter is gratefully acknowledged.

the question of what sizes and configurations of such networks of collaboration are likely to emerge. In some models of natural systems, we have already seen that the networks of interaction that form (i.e., the dynamics of cluster formation) can be predicted ahead of time to have a particular form. For example the Bak-Sneppen model of the coevolution of the species within an evolving ecosystem (described in Chapter 1) gives rise to clusters of coevolution that tend to a power law distribution of cluster size. Clusters of burning trees (forest fires) also show such power law effects, which turn out to be similar to the distribution of casualties in war (Chapters 2 and 3). Clustering of force units in the ISAAC "distillation" model of manoeuvre warfare produces fractal clustering (Chapter 4).

We wish to finish with an extended example analysis of the ISAAC distillation; recall that this is a simple agent-based model of land warfare, incorporating small rule sets that govern agent decisionmaking, movement, and engagement. More detail is available in [1]. Figure 5.1 is a screenshot of the start of a typical ISAAC simulation run.

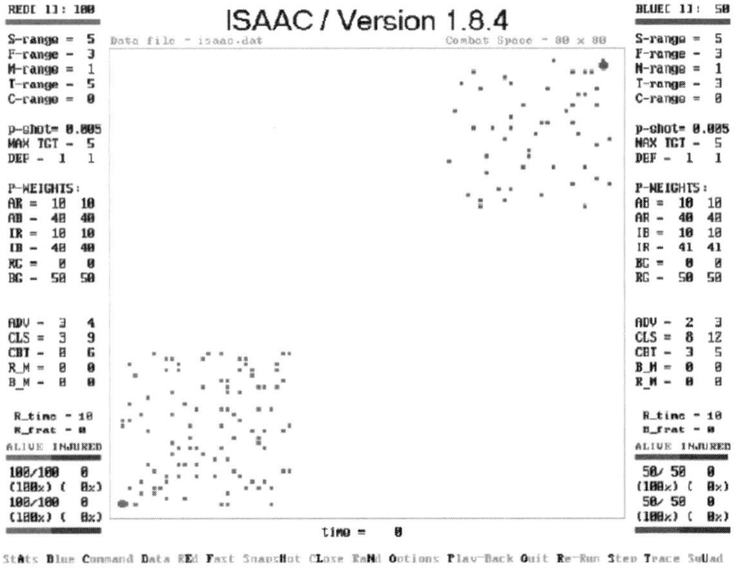

Figure 5.1: Screenshot of the Start of a Typical ISAAC Simulation Run

ISAAC is based on cellular automata, which uses simple local rules to describe interactions between units. Different scenarios can be set up by changing the parameters of these rules. The initial laydown of forces is carried out stochastically, by the model within user-defined limits, and so different replications of the same basic scenario are possible.

The particular ISAAC scenario we use in this chapter was supplied to us by Dr. Gary Horne of the U.S. Marine Corps Warfighting Lab and has three phases. In the first phase, the Red forces move to meet the Blue; the second phase is the engagement between Red and Blue forces. It is in the third phase that either the Red forces take control of the Blue flag (in the top right corner) or the Blue forces retain control of their flag. This scenario is interesting as in all stochastic replications but one (replication 40), the Red forces are successful in achieving their goal of taking control of the Blue flag. The analysis that follows explores the clustering and "swarming" of the agents and the similarities and differences in the replications.

Clustering and Swarming

In order to analyse the swarming dynamics of cluster formation and dissolution in ISAAC, we need to consider firstly what this means. There are two ways to define a cluster of agents. The first, and most usual, is to define neighbouring agents only by those that are north, south, east, or west adjacent to the agent in question, known as *nearest neighbour clustering*. The second (and although most intuitive, less used) definition is to include all eight neighbours of the central agent as part of a cluster, as shown in Figure 5.2, which is known as *next nearest neighbour clustering*.

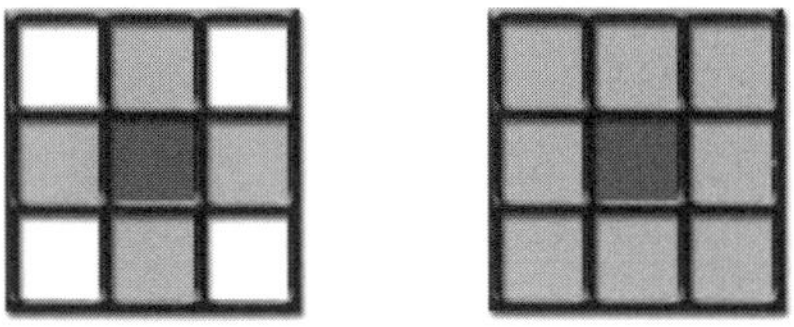

Figure 5.2: Nearest and Next Nearest Neighbour Clustering

For a single timestep in any run of a cellular automata-based model, the number and size of clusters of agents can be determined using a simple algorithm–the Hoshen-Kopelman algorithm–for the nearest neighbour case. This algorithm can be modified for use with eight neighbours. For details of the algorithm, see references [2,3].

Cluster Distribution

Once the cluster numbers and sizes can be determined, there are a number of ways to analyse the data. The first that we look at is the size of the largest cluster. This gives an indication of the ability of the agents to cluster or the amount of dispersal of the agents. For example, if the largest cluster size is near to the total number of agents, we know that that is the only cluster. However, if the largest cluster is small, then we know that the agents are dispersed in many small clusters.

The following plots are of the largest cluster size against the timestep for several different replications with different random seeds for the same basic run of the ISAAC model. The clustering algorithm used is that of nearest neighbours and different plots are graphed for Red and Blue agents. The agents can be ordered by state–alive, injured, or dead. The plots that follow are for only those agents that are alive. For each timestep, we plot the largest cluster size for Blue and the largest cluster size for Red. For the first iteration of our example run of ISAAC, Figure 5.3 shows the evolution of the largest cluster size for Red as a function of simulated time. Figure 5.4 shows the same thing for Blue.

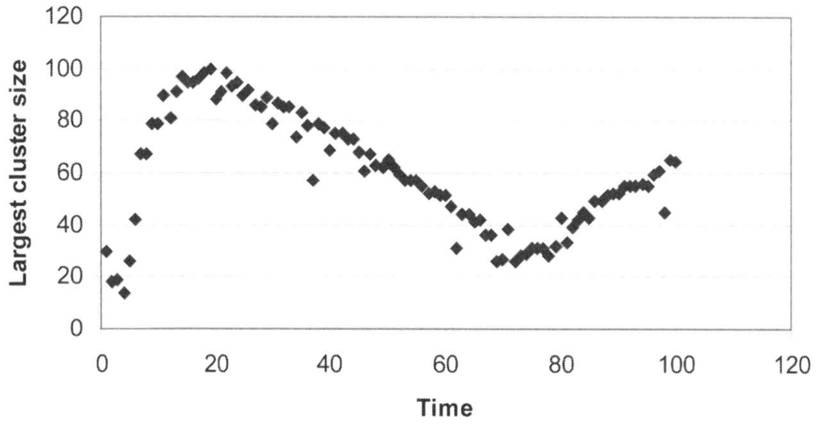

Figure 5.3: Largest Cluster Size as a Function of Simulated Time (First Iteration, Red Agents)

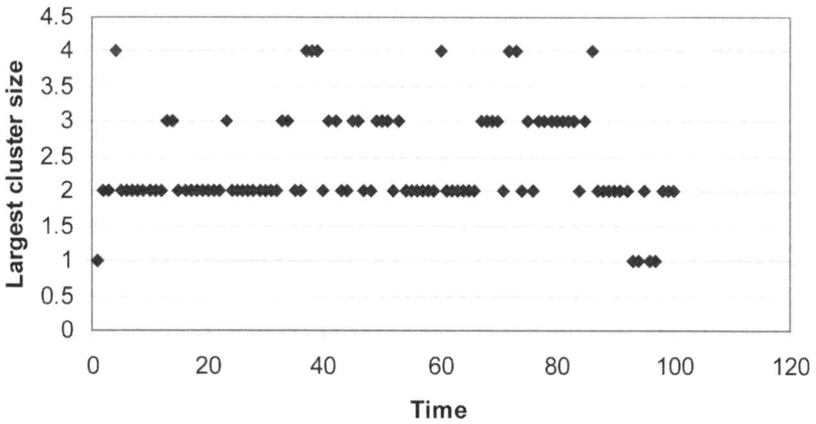

Figure 5.4: Largest Cluster Size as a Function of Simulated Time (First Iteration, Blue Agents)

In these two plots (Figures 5.3 and 5.4), it is possible to see a smoothly changing pattern in the largest cluster size for the Red forces, but not for the Blue. Looking at the clustering of

Red, it is clear that the dynamic behaviour can be split into distinct areas in time by the rate of change of cluster size–the slope of the plot. The distinct areas in time of the changes in the slope of the plot of largest cluster size of Red agents correspond to the three phases of the ISAAC run.

In further replications of the ISAAC run, a similar pattern emerges–each time Red succeeding in his objective. However, in the 40th replication, Red fails to secure the Blue flag. Dr. Horne estimates that the Red force is successful in excess of 100 replications, with only this one failure. Figures 5.5 and 5.6 show the evolution of the largest cluster size for this replication.

We can see that there is now no clear evidence of a third phase of operation in the plot of the largest Red cluster size. In fact, the plot of the largest Blue cluster size is now more structured and shows evidence of a third phase of the force's operation. We suggest that Red's failure is due to the increased clustering ability of the Blue forces, thereby reaching a greater largest cluster size than in other replications. Such behaviour is consistent with the mathematical metamodel of ISAAC discussed in Chapter 4. This metamodel indicates that agile clustering and reclustering should lead to better local force ratios and hence improved ability to cause attrition to the enemy (locally) and to thus move freely.

To gain some additional insight, let us now consider the largest cluster size at each timestep of the simulation, and plot this as a frequency distribution of cluster size. We have done this for a number of replications. Figure 5.7 shows the frequency of largest cluster size for Red agents for typical and exceptional replications.

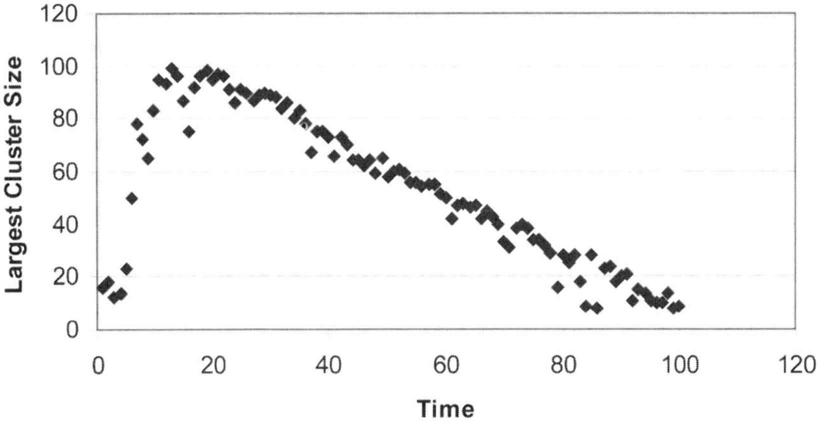

Figure 5.5: Largest Cluster Size as a Function of Time (40th Iteration, Red)

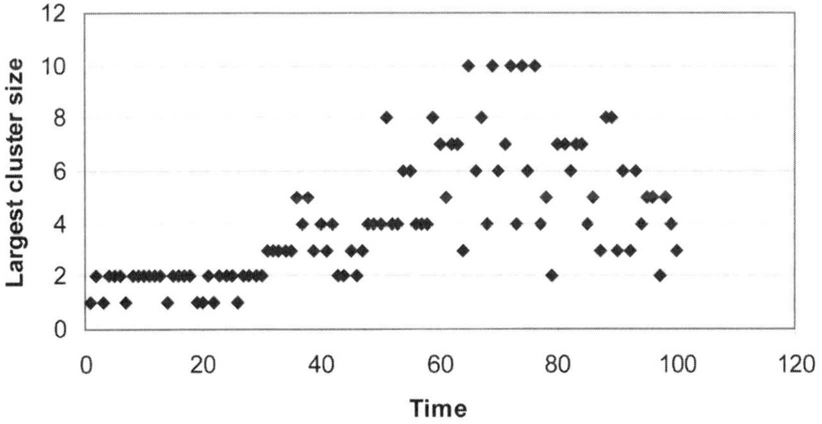

Figure 5.6: Largest Cluster Size as a Function of Time (40th Iteration, Blue)

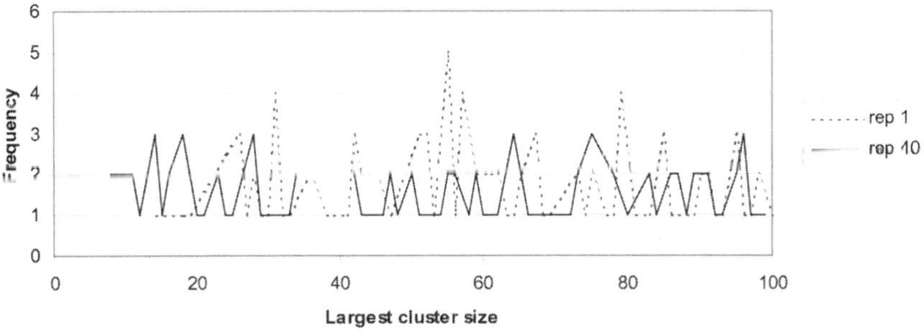

Figure 5.7: Frequency Distribution of the Largest Cluster Size for Red Agents

We can see from this that Red is able to generate a wide spectrum of cluster sizes. Figure 5.8 shows the same plot for the Blue agents with the number of each replication shown on the plot.

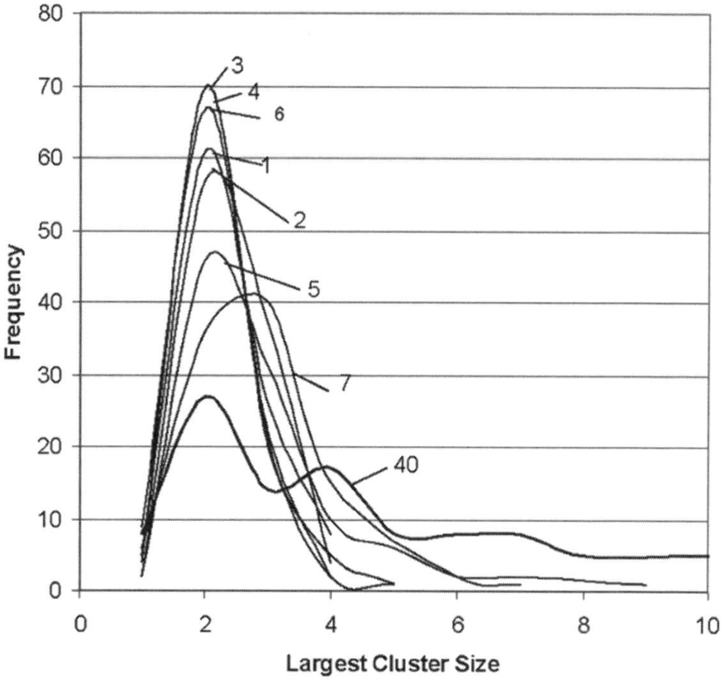

Figure 5.8: Frequency Distribution of the Largest Cluster Size for Blue Agents

We can see from Figure 5.8 that the spread of clusters for Blue is much smaller in general. However, for the 40th replication, Blue is able to generate a wider spread of clusters, and thus succeed.

The Distribution of Cluster Size

Let us move on from largest cluster size now and look at all the clusters formed over time in the simulation. From our mathematical metamodel of ISAAC discussed in Chapter 4, and from the general emergent behaviour of natural systems we have discussed in this book, we anticipate that the distribution of cluster size should approximate to a power law distribution. Thus on a log-log scale, the distribution of cluster size should be a straight line, with end effects where the assumptions break down.

Firstly let us just look at one replication of the simulation. Figure 5.9 shows the distribution of cluster size for Red agents for the 2nd replication of ISAAC, plotted on a log-log scale.

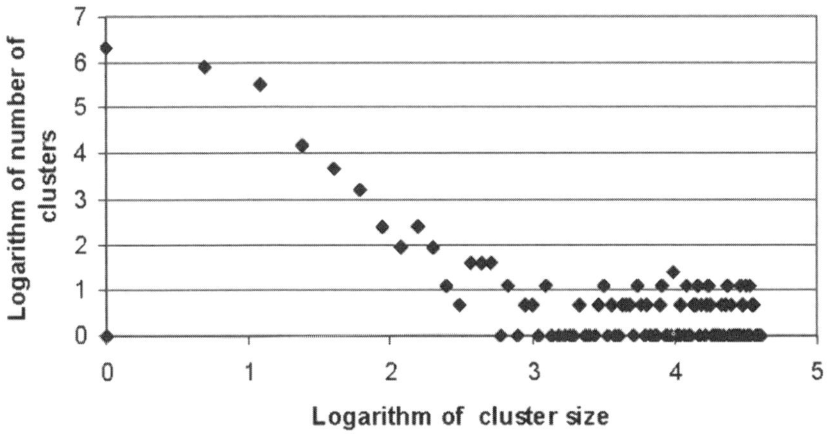

Figure 5.9: Distribution of Cluster Sizes (2nd Replication, Red Agents)

We can see that in the intermediate regime, the plot forms a straight line, confirming the theoretical expectation. Figure 5.10 shows a similar plot with the other replications superimposed.

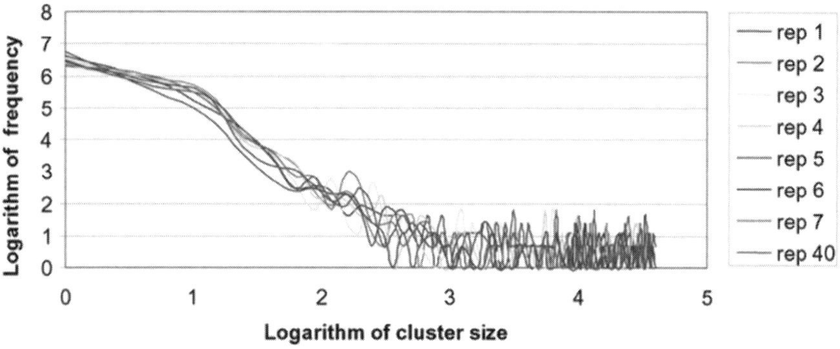

Figure 5.10: Distribution of Cluster Size for Red Agents

Finally, from theory and from the analysis in Chapter 3, we expect that the time series of casualties produced by a model such as ISAAC should show evidence of fractal clustering in time. In precise terms, this implies that the power spectrum of the time series of casualties should be related to the casualty size as a power law relationship. Using a related distillation model called MANA [4], evidence of this effect has been found by Lauren [5], as we have already discussed in Chapter 3. This is an area that we intend to investigate further in the context of building metamodelling equations.

FINAL THOUGHTS

We started by considering what we can learn from natural systems: an ecosystem in which species coevolve locally; a fluid forming an interface when it is pinned; the effect of forest fires. All of these show regularities and emergent behaviours of the whole system that can be captured and deduced using mathematical models. We have also shown how the same ideas of local coevolution within such "open" systems are very relevant to thinking about the consequences of a network-centric form of warfare, where units coevolve (self-synchronise) across an information grid. By exploiting this linkage, it is possible to build quantitative models that help us to understand the likely emergent behaviour of such coevolving networks of force interaction.

This is work in progress that we hope will contribute to the new science of understanding, analysing, and modelling the effects of Information Age warfare. In doing so, we aim, as remarked in the preface to a previous contribution,[2] to "gain a deeper understanding not only of conflict, but also of the avoidance of conflict, which is the ultimate aim of the political/military art."

REFERENCES

1 ILLACHINSKI A (2000). "Irreducible Semi-Autonomous Adaptive Combat (ISAAC): An Artificial Life Approach to Land Warfare." *Military Operations Research.* Vol 5 No 3. pp. 29-46.

2 HOSHEN J and KOPELMAN R (1976). "Percolation and cluster distribution. I. Cluster multiple labeling technique and critical concentration algorithm." *Phys. Rev.* B14, p. 3438.

3 http://www.splorg.org/people/tobin/projects/hoshenkopelman/hoshenkopelman.html (Aug 1, 2003)

4 LAUREN M K and STEPHEN R T (2002). "Map-Aware Non-Uniform Automata (MANA)–A New Zealand Approach to Scenario Modelling." *Journal of Battlefield Technology.* Vol 5 No 1. pp. 27-31.

5 LAUREN M K and STEPHEN R T (2002). "Fractals and Combat Modelling: Using MANA to Explore the Role of Entropy in Complexity Science." Paper prepared for *Fractals.* Defence Technology Agency. Auckland, New Zealand.

---

[2] Moffat J. *Command and Control in the Information Age; Representing its Impact.* The Stationery Office. London, UK, 2002.

APPENDIX

OPTIMAL CONTROL WITH A UNIQUE CONTROL SOLUTION

In this Appendix, we investigate the case of a unique optimal control solution to the problem of system control, and show that this unique solution is of the form of *bang-bang* control when the system is linear in nature.

As in Chapter 1, we assume that our system can be described by the functional relationship:

$$\frac{dX_i}{dt} = F_i\left(X_1(t),\ldots X_n(t);\lambda_1(t),\ldots\lambda_m(t)\right),\ (i=1,\ldots.n)$$

where F_i are the rate laws, and $\lambda_i(t)$ are the control variables. In matrix/vector notation, we write this as:

$$\dot{X}(t) = F(X(t),\lambda(t))$$

with initial conditions $X_i(t_0) = X_i^0$.

We assume fairly weak conditions on the continuity of F, X, and λ sufficient to make the equations "well behaved."

We consider here analysis of this relationship as time varies over a fixed time interval $[t_0,t_1]$. At time t_1, the state variables will have values $X_i(t_1)$, and the objective is to maximise or minimise a linear combination of these endstate values. The problem can thus be written as:

$$\text{Optimise } \sum_{i=1}^{n} c_i X_i(t_1)$$

(where the c_i are constant coefficients or "weights") subject to the constraints:

$$\dot{X}(t) = F(X(t), \lambda(t))$$

$$X_i(t_0) = X_i^0 \quad \forall i.$$

We now define a process [1] that yields a necessary condition for a vector of control variables $\lambda(t)$ to optimise the objective function. In other words, any control vector that gives rise to an optimal value of the objective function must satisfy this condition. Although this does not guarantee that a solution $\lambda(t)$ satisfying this condition is optimal, other information (such as the uniqueness of such a solution) can be used in particular cases to prove that $\lambda(t)$ is indeed an optimal control vector.

The first step in this procedure is to introduce a set of "dual" variables:

$$\psi_1(t), \ldots\ldots, \psi_n(t)$$

that are defined by the relationships:

$$\dot{\psi}_i(t) = -\sum_{j=1}^{n} \psi_j(t) \frac{\partial F_j(X,\lambda)}{\partial X_i}$$

with final conditions:

$$\psi_i(t_1) = -c_i \quad \forall i$$

where the values c_i are the same as the coefficients appearing in the objective function.

A Hamiltonian function H is now defined by:

$$H(\psi, X, \lambda) = \langle \psi, \dot{X} \rangle$$

where < > denotes the inner product of the two vectors ψ and λ.

Thus:

$$H(\psi, X, \lambda) = \sum_{i=1}^{n} \psi_i(t)\dot{X}_i(t)$$
$$= \sum_{i=1}^{n} \psi_i(t) F_i(X, \lambda)$$

From the definition of H, we have the dual relationship:

$$\dot{X}_i = \frac{\partial H}{\partial \psi_i} \quad \forall i$$
$$\dot{\psi}_i = -\frac{\partial H}{\partial X_i} \quad \forall i$$

The corresponding "boundary conditions" are:

$$X_i(t_0) = X_i^0 \quad \forall i$$
$$\psi_i(t_1) = -c_i \quad \forall i$$

Pontryagin's Maximum Principle

Using the notation we have now developed, Pontryagin's Maximum Principle [2] states that if $\lambda^*(t)$ is a control vector that maximises (resp. minimises) the objective function:

$$\sum_{i=1}^{n} c_i X_i(t_1)$$

then the Hamiltonian $H(\psi, X, \lambda)$ achieves a minimum (resp. maximum) at the point $\lambda^*(t)$ for any value of X or ψ.

It is worth noting here that if the set U of admissible control vectors is a topologically compact set, then the continuous function H will realise its minimum or maximum value on U.

The control vectors λ that minimise or maximise the Hamiltonian H are called *extremal controls*. If we denote this subset of U by U^*, then we know from Pontryagin's Maximum Principle that if λ^* is an optimal control vector, then $\lambda^* \in U^*$. Thus, if we know that:

1. An optimal control vector exists; and
2. There is only one extremal control (i.e., U^* is a single point)

then the single element of U^* must be the optimal control vector. Whether (1) and (2) apply depends on the particular problem under study.

Determining the Extremal Controls

The process for determining these extremal controls is as follows (for definiteness, we assume that we are maximising the objective function, and hence minimising the Hamiltonian):

1. Compute the form of the Hamiltonian function $H(\psi, X, \lambda)$.

2. Compute the values of λ which minimise $H(\psi, X, \lambda)$ for $\lambda \in U$.

   These are the extremal controls. They are expressed as functions of X and ψ

$$\lambda = \lambda(X, \psi).$$

3. Substitute this extremal control λ into the relations:

$$\dot{X}_i = \frac{\partial H}{\partial \psi_i}$$

$$\dot{\psi}_i = -\frac{\partial H}{\partial X_i}$$

   in order to solve for the extremal system trajectory X^* and extremal dual function ψ^*.

4. Substitute these values into the expression for the extremal control:

$$\lambda = \lambda(X^*, \psi^*)$$

   to give an explicit formulation of this extremal control vector.

   Knowledge of the extremal trajectory X^* allows the objective function:

$$\sum_{i=1}^{n} c_i X_i^*(t_1)$$

   to be evaluated.

LINEAR MODELS

A linear system model is defined as having a relationship of the form:

$$\dot{X}(t) = A(t)X(t) + B(t)\lambda(t) + g(t)$$

in a matrix and vector representation, with initial conditions $X_i(t_0) = X_i^0 \quad \forall i$.

When the system behaviour is of this form, it is possible to characterise the nature of the optimal controls under fairly general conditions. This is particularly the case if the objective function itself is linear, i.e. it is of the form:

$$\int_{t_0}^{t_1} < S, X > + < W, \lambda >$$

where, as before, $< \ >$ denotes the inner product of two vectors, and S and W are time-dependent vectors of known value.

Assume then that the objective function is of this form, and that the system model is linear in the way that we have described. Without loss of generality, we can set $g(t)=0$ and write the system behaviour model in the form:

$$\dot{X} = AX + B\lambda$$

where X is the vector of state variables (e.g., force levels) and λ is the vector of control variables.

Make the transformation:

$$X_{n+1}(t) = \int_{t_0}^{t} < S, X > + < W, \lambda > .$$

The objective function now becomes:

$$\text{Optimise } X_{n+1}(t_1)$$

and the relationship $\dot{X}_{n+1}(t) = < S(t), X(t) > + < W(t), \lambda(t) >$ is added to the set of equations describing the system behaviour. (This can be done since the above equation is linear and so has the same form as the others.)

Consider now the form of the Hamiltonian for such a linear system. We have:

$$\begin{aligned} H(\psi, X, \lambda) &= < \psi, \dot{X} > \\ &= < \psi, AX + B\lambda > \\ &= < \psi, AX > + < \psi, B\lambda > \end{aligned}$$

Since we are interested in the extremal controls λ that maximise or minimise H, only the second term is of interest, the first not being a function of λ.

Let us look at this second term in more detail. We have:

$$< \psi, B\lambda > = \sum_i \psi_i \sum_j B_{ij}\lambda_j = \sum_j (\sum_i \psi_i B_{ij})\lambda_j$$

Let: $$\varphi_j = \sum_i \psi_i B_{ij} .$$

Then: $$< \psi, B\lambda > = \sum_j \varphi_j \lambda_j = < \varphi, \lambda > .$$

Let us assume that the objective function is to be maximised. By Pontryagin's Maximum Principle [2], we thus wish to consider control vectors λ that minimise the Hamiltonian H. This is then equivalent, as we have seen, to minimising the expression $\sum_j \varphi_j(t)\lambda_j(t)$.

We can define such an extremal control vector λ^* as follows, (provided that the set U of all possible control vectors is topologically compact):

$$\text{If } \varphi_j(t) \le 0 \text{ let } \lambda_j^*(t) = \max\{\lambda_j(t), \lambda \in U\}$$
$$\text{and if } \varphi_j(t) > 0 \text{ let } \lambda_j^*(t) = \min\{\lambda_j(t), \lambda \in U\}$$

These are well defined since a continuous function will attain its max or min on a compact set.

For any t in the interval $[t_0, t_1]$ we then have:

$$\sum_j \varphi_j(t)\lambda_j^*(t) \le \sum_j \varphi_j(t)\lambda_j(t).$$

For any control vector λ in the admissible set U of control vectors.

Uniqueness of the Extremal Control for a Linear System

If V is any other extremal control vector, then since it minimises the Hamiltonian H, it must satisfy:

$$\sum_j \varphi_j(t)V_j(t) \le \sum_j \varphi_j(t)\lambda_j(t)$$

for any λ in the admissible set U of control vectors.

However, if $\varphi_j(t) > 0$, then it is clear that:

$$V_j(t) = \min\{\lambda_j(t), \lambda \in U\}.$$

Otherwise it would be possible to define a vector giving a smaller value of the Hamiltonian, contradicting the extremal nature of V. It follows that every extremal vector must be of the form λ^*.

At the points where $\varphi_j(t)$ changes sign, $\lambda_j^*(t)$ changes from an extreme minimum value to an extreme maximum, or vice versa. Such a form of control is known as *bang-bang* since the value "bangs" from one extreme possible value to another, and never assumes any intermediate values. What we have shown is that for a linear system with a linear objective function, every extremal control (including therefore the optimal control) must be in *bang-bang* form. In this sense, the optimal control vector always lies on the boundary of the admissible set U.

If it can be shown that the φ_j are unique, then the above construction yields a unique control vector that must then be the optimal control.[1] Now, we have that:

$$\varphi_j = \sum_i \psi_i B_{ij} .$$

Thus the uniqueness of φ_j depends on the uniqueness of ψ_i $(1 \le i \le n)$.

We have that:

$$\begin{aligned}
\dot{\psi}_i &= -\sum_j \psi_j \frac{\partial}{\partial X_i}(\dot{X}_j) \\
&= -\sum_j \psi_j \frac{\partial}{\partial X_i}\left(\sum_l A_{jl} X_l + \sum_m B_{jm} \lambda_m\right) \\
&= -\sum_j \psi_j A_{ji}
\end{aligned}$$

with final conditions:

$$\psi(t_1) = -(c_1, \ldots\ldots, c_{n+1}) = -(0, \ldots\ldots, 0, 1) .$$

[1] It can be shown that for this type of system, an optimal control must exist [1] [2].

Since the matrix A is known, this equation has a unique solution [1] and thus the vector ψ is unique. Hence, the optimal control for such a linear system that optimises the objective function is precisely defined by the *bang-bang* control function λ^*.

REFERENCES

1 CONNORS M M and TEICHROEW D (1967). *Optimal Control of Dynamic Operations Research Models.* International Textbook Co. Pennsylvania, USA.

2 ROZONOER L T (1959). L.S. *Pontryagin's Maximum Function Principle in its Application to the Theory of Optimum Systems–I, II, III.* Avtomatika i Telemikhanika 20. p. 1320 et seq. Translated in the journal Automation and Remote Control (1959). 20. p. 1288 et seq.

ABOUT THE AUTHOR

Professor James Moffat is a Senior Fellow of the Defence Science and Technology Laboratory (Dstl), UK, a Fellow of Operational Research, a Fellow of the Institute of Mathematics and its Applications, and a visiting Professor at Cranfield University, UK. He was awarded the President's Medal of the Operational Research Society in the year 2000. He holds a first class honours degree and a Ph.D. in Mathematics, and was awarded the Napier Medal in Mathematics by Edinburgh University. He has worked for the past 20 years or so on defence-related operational analysis problems and aerospace technology research. His current research interest is in building analysis tools and models that capture the key effects of human decisionmaking and the other aspects of C4ISR.

Catalog of CCRP Publications

(* denotes a title no longer available in print)

Coalition Command and Control*
(Maurer, 1994)

Peace operations differ in significant ways from traditional combat missions. As a result of these unique characteristics, command arrangements become far more complex. The stress on command and control arrangements and systems is further exacerbated by the mission's increased political sensitivity.

The Mesh and the Net
(Libicki, 1994)

Considers the continuous revolution in information technology as it can be applied to warfare in terms of capturing more information (mesh) and how people and their machines can be connected (net).

Command Arrangements for Peace Operations
(Alberts & Hayes, 1995)

By almost any measure, the U.S. experience shows that traditional C2 concepts, approaches, and doctrine are not particularly well suited for peace operations. This book (1) explores the reasons for this, (2) examines alternative command arrangement approaches, and (3) describes the attributes of effective command arrangements.

Standards: The Rough Road to the Common Byte
(Libicki, 1995)

The inability of computers to "talk" to one another is a major problem, especially for today's high technology military forces. This study by the Center for Advanced Command Concepts and Technology looks at the growing but confusing body of information technology standards. Among other problems, it discovers a persistent divergence between the perspectives of the commercial user and those of the government.

What Is Information Warfare?*
(Libicki, 1995)

Is Information Warfare a nascent, perhaps embryonic art, or simply the newest version of a time-honored feature of warfare? Is it a new form of conflict that owes its existence to the burgeoning global information infrastructure, or an old one whose origin lies in the wetware of the human brain but has been given new life by the Information Age? Is it a unified field or opportunistic assemblage?

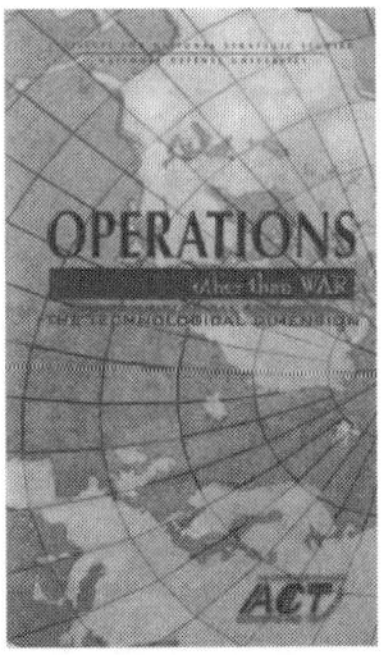

Operations Other Than War*
(Alberts & Hayes, 1995)

This report documents the fourth in a series of workshops and roundtables organized by the INSS Center for Advanced Concepts and Technology (ACT). The workshop sought insights into the process of determining what technologies are required for OOTW. The group also examined the complexities of introducing relevant technologies and discussed general and specific OOTW technologies and devices.

Dominant Battlespace Knowledge
(Johnson & Libicki, 1996)

The papers collected here address the most critical aspects of that problem—to wit: If the United States develops the means to acquire dominant battlespace knowledge, how might that affect the way it goes to war, the circumstances under which force can and will be used, the purposes for its employment, and the resulting alterations of the global geomilitary environment?

Interagency and Political-Military Dimensions of Peace Operations: Haiti - A Case Study
(Hayes & Wheatley, 1996)

This report documents the fifth in a series of workshops and roundtables organized by the INSS Center for Advanced Concepts and Technology (ACT). Widely regarded as an operation that "went right," Haiti offered an opportunity to explore interagency relations in an operation close to home that had high visibility and a greater degree of interagency civilian-military coordination and planning than the other operations examined to date.

The Unintended Consequences of the Information Age*
(Alberts, 1996)

The purpose of this analysis is to identify a strategy for introducing and using Information Age technologies that accomplishes two things: first, the identification and avoidance of adverse unintended consequences associated with the introduction and utilization of information technologies; and second, the ability to recognize and capitalize on unexpected opportunities.

Joint Training for Information Managers*
(Maxwell, 1996)

This book proposes new ideas about joint training for information managers over Command, Control, Communications, Computers, and Intelligence (C4I) tactical and strategic levels. It suggests a substantially new way to approach the training of future communicators, grounding its argument in the realities of the fast-moving C4I technology.

Defensive Information Warfare*
(Alberts, 1996)

This overview of defensive information warfare is the result of an effort, undertaken at the request of the Deputy Secretary of Defense, to provide background material to participants in a series of interagency meetings to explore the nature of the problem and to identify areas of potential collaboration.

Command, Control, and the Common Defense
(Allard, 1996)

The author provides an unparalleled basis for assessing where we are and were we must go if we are to solve the joint and combined command and control challenges facing the U.S. military as it transitions into the 21st century.

Shock & Awe: Achieving Rapid Dominance*

(Ullman & Wade, 1996)

The purpose of this book is to explore alternative concepts for structuring mission capability packages around which future U. S. military forces might be configured.

Information Age Anthology: Volume I*

(Alberts & Papp, 1997)

In this first volume, we will examine some of the broader issues of the Information Age: what the Information Age is; how it affects commerce, business, and service; what it means for the government and the military; and how it affects international actors and the international system.

Complexity, Global Politics, and National Security

(Alberts & Czerwinski, 1997)

The charge given by the President of the National Defense University and RAND leadership was threefold: (1) push the envelope; (2) emphasize the policy and strategic dimensions of national defense with the implications for Complexity Theory; and (3) get the best talent available in academe.

Target Bosnia: Integrating Information Activities in Peace Operations*
(Siegel, 1998)

This book examines the place of PI and PSYOP in peace operations through the prism of NATO operations in Bosnia-Herzegovina.

Coping with the Bounds
(Czerwinski, 1998)

The theme of this work is that conventional, or linear, analysis alone is not sufficient to cope with today's and tomorrow's problems, just as it was not capable of solving yesterday's. Its aim is to convince us to augment our efforts with nonlinear insights, and its hope is to provide a basic understanding of what that involves.

Information Warfare and International Law*
(Greenberg, Goodman, & Soo Hoo, 1998)

The authors, members of the Project on Information Technology and International Security at Stanford University's Center for International Security and Arms Control, have surfaced and explored some profound issues that will shape the legal context within which information warfare may be waged and national information power exerted in the coming years.

Lessons From Bosnia: The IFOR Experience*
(Wentz, 1998)

This book tells the story of the challenges faced and innovative actions taken by NATO and U.S. personnel to ensure that IFOR and Operation Joint Endeavor were military successes. A coherent C4ISR lessons learned story has been pieced together from firsthand experiences, interviews of key personnel, focused research, and analysis of lessons learned reports provided to the National Defense University team.

Doing Windows: Non-Traditional Military Responses to Complex Emergencies
(Hayes & Sands, 1999)

This book provides the final results of a project sponsored by the Joint Warfare Analysis Center. Our primary objective in this project was to examine how military operations can support the long-term objective of achieving civil stability and durable peace in states embroiled in complex emergencies.

Network Centric Warfare
(Alberts, Garstka, & Stein, 1999)

It is hoped that this book will contribute to the preparations for NCW in two ways. First, by articulating the nature of the characteristics of Network Centric Warfare. Second, by suggesting a process for developing mission capability packages designed to transform NCW concepts into operational capabilities.

Behind the Wizard's Curtain
(Krygiel, 1999)

There is still much to do and more to learn and understand about developing and fielding an effective and durable infostructure as a foundation for the 21st century. Without successfully fielding systems of systems, we will not be able to implement emerging concepts in adaptive and agile command and control, nor will we reap the potential benefits of Network Centric Warfare.

Confrontation Analysis: How to Win Operations Other Than War
(Howard, 1999)

A peace operations campaign (or operation other than war) should be seen as a linked sequence of confrontations, in contrast to a traditional, warfighting campaign, which is a linked sequence of battles. The objective in each confrontation is to bring about certain "compliant" behavior on the part of other parties, until in the end the campaign objective is reached. This is a state of sufficient compliance to enable the military to leave the theater.

Information Campaigns for Peace Operations
(Avruch, Narel, & Siegel, 2000)

In its broadest sense, this report asks whether the notion of struggles for control over information identifiable in situations of conflict also has relevance for situations of third-party conflict management—for peace operations.

Information Age Anthology: Volume II*

(Alberts & Papp, 2000)

Is the Information Age bringing with it new challenges and threats, and if so, what are they? What sorts of dangers will these challenges and threats present? From where will they (and do they) come? Is information warfare a reality? This publication, Volume II of the Information Age Anthology, explores these questions and provides preliminary answers to some of them.

Information Age Anthology: Volume III

(Alberts & Papp, 2001)

In what ways will wars and the military that fight them be different in the Information Age than in earlier ages? What will this mean for the U.S. military? In this third volume of the Information Age Anthology, we turn finally to the task of exploring answers to these simply stated, but vexing questions that provided the impetus for the first two volumes of the Information Age Anthology.

Understanding Information Age Warfare

(Alberts, Garstka, Hayes, & Signori, 2001)

This book presents an alternative to the deterministic and linear strategies of the planning modernization that are now an artifact of the Industrial Age. The approach being advocated here begins with the premise that adaptation to the Information Age centers around the ability of an organization or an individual to utilize information.

Information Age Transformation
(Alberts, 2002)

This book is the first in a new series of CCRP books that will focus on the Information Age transformation of the Department of Defense. Accordingly, it deals with the issues associated with a very large governmental institution, a set of formidable impediments, both internal and external, and the nature of the changes being brought about by Information Age concepts and technologies.

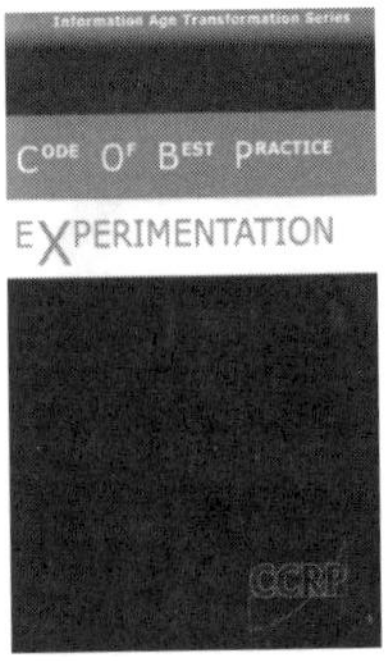

Code of Best Practice for Experimentation
(CCRP, 2002)

Experimentation is the lynch pin in the DoD's strategy for transformation. Without a properly focused, well-balanced, rigorously designed, and expertly conducted program of experimentation, the DoD will not be able to take full advantage of the opportunities that Information Age concepts and technologies offer.

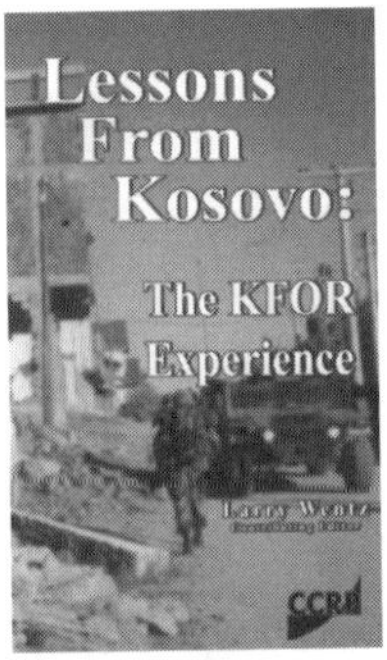

Lessons From Kosovo: The KFOR Experience
(Wentz, 2002)

Kosovo offered another unique opportunity for CCRP to conduct additional coalition C4ISR-focused research in the areas of coalition command and control, civil-military cooperation, information assurance, C4ISR interoperability, and information operations.

NATO Code of Best Practice for C2 Assessment
(2002)

To the extent that they can be achieved, significantly reduced levels of fog and friction offer an opportunity for the military to develop new concepts of operations, new organisational forms, and new approaches to command and control, as well as to the processes that support it. Analysts will be increasingly called upon to work in this new conceptual dimension in order to examine the impact of new information-related capabilities coupled with new ways of organising and operating.

Effects Based Operations
(Smith, 2003)

This third book of the Information Age Transformation Series speaks directly to what we are trying to accomplish on the "fields of battle" and argues for changes in the way we decide what effects we want to achieve and what means we will use to achieve them.

The Big Issue
(Potts, 2003)

This Occasional considers command and combat in the Information Age. It is an issue that takes us into the realms of the unknown. Defence thinkers everywhere are searching forward for the science and alchemy that will deliver operational success.

Power to the Edge: Command...Control... in the Information Age

(Alberts & Hayes, 2003)

Power to the Edge articulates the principles being used to provide the ubiquitous, secure, wideband network that people will trust and use, populate with high quality information, and use to develop shared awareness, collaborate effectively, and synchronize their actions.